THE RACE FOR TECHNOLOGY

Conquering The High Frontier

Thomas W. Becker

authorHOUSE®

AuthorHouse™
1663 Liberty Drive, Suite 200
Bloomington, IN 47403
www.authorhouse.com
Phone: 1-800-839-8640

First published by AuthorHouse   8/25/2008

ISBN: 978-1-4389-0937-0 (sc)

Printed in the United States of America
Bloomington, Indiana

This book is printed on acid-free paper.

All photos, diagrams and charts in the book are the original work of the author unless indicated otherwise.

Books by the same author

Pageant Of World Commemorative Coins. Racine, Wisconsin: Whitman Publishing Company, 1962.

The Coin Makers. Garden City: Doubleday and Company, 1969.

EISENHOWER The Man, The Dollar and The Stamps. The American Mint and Postal Society and Mintmaster Inc., 1971.

Exploring Tomorrow In Space. Garden City: Sterling Publishing Company Inc., 1972
(Foreword by Dr. Wernher von Braun).

Our American Coins. The U.S. Treasury Dept (on contract), Bureau Of The Mint, Washington DC, 1972.

Aerospace: Crossing The Space Frontier. University of Missouri, Center for Distance and Independent Study, Columbia, Mo. 1988, rev 1998. Gifted high school self-study course in the history of space technology 1920-present. Mid-term and Final exams.

Studying Planet Earth: The Satellite Connection. University of Missouri, Center for Distance and Independent Study, Columbia, Mo. 1997. Gifted high school self-study course in remote sensing and Earth studies. Mid-term and Final exams.

Eight Against The World: Warriors Of The Scientific Revolution. Bloomington, Indiana: Author House Publisher, 2007.

A Season Of Madness: Life And Death In The 1960s. Bloomington, Indiana: Author House Publisher, 2007.

The Race For Technology: Conquering The High Frontier. Bloomington, Indiana: Author House Publisher, 2008.

Book 3 of a Technology Trilogy

For Barbara, Paul, Stephen, David and Marilyn,
with many thanks for so many things, this book
is inscribed with deepest affection.

"The woods are lovely, dark and deep,
But I have promises to keep,
And miles to go before I sleep. . ."

Robert Frost

For my friends in Europe:

"I have eaten your bread and salt, I have drunk your water and wine,
The deaths ye died I have watched beside and
the lives that ye led were mine.

Was there aught that I did not know, in vigil, toil or ease,
One joy or woe that I did not know, dear hearts across the seas.

I have written a tale of our lives, for a sheltered people's mirth,
In jesting guise but ye are wise and ye know what the jest is worth."

Rudyard Kipling

On The Cover: In the early morning hours at Launch Pad 39A, Cape
Canaveral, the Apollo 14 spacecraft atop its Saturn rocket awaits last
minute check-out before lift-off in 1971. The Apollo moon-craft will
carry three astronauts to the moon during America's breathtaking era
of *technology-on-demand* in the race for supremacy at the cutting edge
of the high frontier.

CONTENTS

PREFACE

T his third book in the Technology Trilogy is about technology once again, written to explain the human side of science and technology caught up in the difficult years of global conflict and politics. Technology is not difficult to understand though at times it may seem to be complicated. Against a background of space exploration, the Cold War and such events as airborne refueling of B52 bombers and the amazing success of the European Space Agency's point blank imaging of Halley's Comet, dramatic events stand out in stark relief. As witness to these events, I wanted to describe how the event took place as much as how the technology was emerging, and to provide the human touch to a book that would appeal to people who simply are curious or who just like to read.

No matter how hard I tried, either overtly or covertly, it was impossible to escape the relentless grip of the tentacles of the Cold War that continued to weave itself into every aspect of the global culture in the last half of the 20th century. The period 1945 – 1999 with all its creative engineering and purposeful technology was America's response to Soviet technological advances and a daily demand on our nation's intellectual and material resources. Although the allies worked ceaselessly to devise new strategies and more advanced mechanical and electronic methods of meeting and keeping watch over the Soviets, the international game of "catch me if you can" was an inescapable contest forcing America to invent newer and newer technologies to keep abreast of Soviet science and technology.

This brutal reality of the 20th century is the foundation of this book in which world events are seen pitted against creative *technology on demand.* Occasionally we were caught off guard as in the sudden extension of the runway at Tyuratam in the Soviet Union, the major launch site for Soviet manned rockets and missions, that was accomplished so subtly it nearly escaped analysis by our intelligence community. Suddenly it was obvious the length of the runway was being increased to accommodate a new type of *aerospace* craft built on a design copied from our own Space Shuttle (see the illustration on page 89 prepared by the US Department of Defense). By the time our intelligence community decided on the reason for the extension, the Soviet *Buran* Space Shuttle was ready to be flight-tested.

At other times, the game of *one-up-man-ship* was so closely matched it became almost comical. A Soviet scientist who defected to America in the 1980s confessed in a security interview that the top priority of Soviet technology spies in America was to learn the ingredients that made up the thermodynamic tiles on the American Space Shuttle. That piece of unrelated information, when put together with the discovery of the Tyuratam runway increase, began to create a larger meaning to the puzzle of one of the Soviet Union's newest crash projects. There were other pieces to that puzzle that are discussed in Chapter 8 Russian Secrets.

As a teacher of space technology, especially in Europe, it was often difficult to determine just how far I should go with my students in teaching about American versus Soviet technologies. The question was finally answered when the US Department of Defense sent me a set of full color art renderings in 1986 of emerging Soviet weapons and space technologies. I felt greatly relieved to learn the pictures had just been released to the military and civilian defense communities in a yearly publication titled *Soviet Military Power* issued by the DOD. That kindness also created an awareness in my own mind of how close I was to the cutting edge of the intelligence community. In a sense, it was something of a spooky realization; I would have to teach my students with very careful wording.

This book approximately covers the years 1971 to about the start of the 3rd Millennium or so – a span of thirty years during which I found myself thrust into an international arena on several fronts and a very complex life style that allowed very little rest and required a constant self-imposed regimen of international studies. During the forty or so years of my technology career, I managed to write three hundred or more popular and journal articles. It was when I began to receive mail from Germany, Poland, Mexico, France etc. asking for copies of my teaching materials that I became uneasy about what I should be sending abroad. Complicated – very complicated! A cursory glance at the Acknowledgements reveals the kinds of organizations I was working with.

What happened to America in these thirty years? The question begs for some kind of informed answer. One answer is an answer, but it is not THE answer.

We were a nation once accustomed to great achievements in those thirty years. We not only have traded creative science for militarism, but over the years we also traded our global leadership position for a new narcissistic position that greatly diluted our cultural value system. The American people seem to have lost their vision for the future because that vision no longer lives in the hearts and minds of our leaders. NASA has slipped into an apparent mediocrity; neither the space station nor the quest for Mars nor the idea of a human colony on the Moon have given Americans the kind of national unity that was our vision in the last half of the 20th century.

There have been no grand projects that have benefited humankind and helped the human race better understand its relationship to the universe. The question still begs for an answer because no one has articulated THE answer to this question. And no leadership in America has come forward to define a more wholesome "national vision" or an attempt to structure a wholesome value system for our culture.

Another possible answer is the fact that America has lost a sense of its own value system. American values at the start of the 3rd millennium are focused on four ideas: money, sex, technology and politics. Hollywood's celebrities have sunk to new lows both on and off the silver screen; America's taste in television is sitcoms and unreality shows. Although quite a few motion pictures have shown creativity, the common level of most is (again) mediocrity. As affluent as America is, the thing that drives our culture today is entertainment – from sports to television to the great wasteland of the Internet. In the overall scheme of things, *how much do we really care* about the public obsession with Britany Spears' underwear – or the absence of it - in public?

The great leap forward in technology in the past fifty years came about by adapting the electromagnetic spectrum to our cultural needs, a piece of physics that was in good use already at the close of World War II. Foreign space programs (notably Japan, Canada and the European Space Agency) have offered partnerships for large and worthwhile projects reaching far out into the cosmos but America cannot seem to form lasting partnering relationships unless we're "in charge."

NASA no longer carries its message to the public with vigor and is afflicted with the dreaded disease known as the NIH syndrome: *Not Invented Here*. Through pride, arrogance or just plain stupidity, America chases a leaderless horse. America no longer creates leaders of powerful vision and national imagination. The Congress virtually destroyed the NASA reputation and capability when it suddenly withdrew funding for manned space missions beyond Apollo 11, because our Congressmen were not astute enough to realize the enormous amounts of commercial spin-offs Americans created during the process of reaching for the Moon. This was a glaring mistake of catastrophic international proportion.

The content of this third book of the trilogy has to do with the race for technology superiority and how the nature of technology advances have carried America into the 3rd millennium. This third book has an international breadth that includes an ever-widening scope from Western Europe and the Soviet Union to emerging new technologies of a global nature. Arthur C. Clarke's *Global Village*, where boundaries between nations have been erased by all-seeing satellites and an extended human presence in space, is a reality!

One of the best short articles on the subject of a national vision is one written for the July 25 1994 issue of TIME magazine by Hugh Sidey titled "Why We Went To the Moon." In that article, Sidey wrote prophetically, *"But our moon legacy leaves a daunting question. Why can we not find such a national project in today's contentious world that would give us a common purpose?"* In short, why does America not nurture the kinds of visionary leaders who can carry the country to the edge of grand achievements to build national unity and pride once more? Individually and cooperatively we must give up the fear of losing what we already possess and once again adopt the kinds of social, political and cultural pursuits that once made America the great hope for the world.

The space frontier is NOT a place. . . it is a concept of the human mind in all its most tantalizing aspects, calling out to us with the promise of great expectations. It is difficult for the public to understand the urgency of maintaining the high ground in space, simply because people not associated with global space development have little or no reference beyond what they see on television and the Internet, or read in various publications. Humans can't understand a concept unless they can relate to it emotionally and personally.

The public can't relate to things they can't see, touch, feel, smell or taste unless these things are constantly in front of them. Americans especially have little or no background understanding of space exploration and development; they react on an emotional level. The space frontier is not personal enough, and it is too technological at a time when we are absolutely inundated daily with hundreds of little different kinds of cultural technologies and issues. We can see the International Space Station floating in space. . . *on television. . .*but it is too far removed from our senses to be personal. Americans cannot buy into a concept, on an emotional level, that they can't believe in – or relate to on a sensory level – or envision its benefits. Remember that human beings, on physical, mental and emotional levels, are enormously complicated.

No matter what else I might be, I am first and foremost a teacher about the world of tomorrow. These three books – the Technology Trilogy – were written to present a view of technology that was begun five hundred years ago by the "natural philosophers" (scientists) of the Renaissance. That technology was created out of the commitment and dedication of a group of men who gave up their very lives for what they believed in. Every century since then has merely introduced refinements of their original work and, through modern inventions, allowed us to discover the new technologies of today. The 20th century experienced an absolute explosion of new technologies; the 21st century will tell us if we can live with the results.

I highly recommend the Book *The Greatest Generation* by newscaster Tom Brokow if you want to understand the time period from WWII onward. It is filled with astonishing real-life examples of the birth of America's value system of the 1940s and 1950s that emphasized the ideals of sacrifice, commitment, responsibility, accountability, and the enormous meaning of *family* as the foundation of American culture. That value system gave way to a new and considerably weakened value system of the 1960s and 1970s that in turn became the nearly invisible value system of the 1980s and 1990s leading to our "me-me-me" cultural values in the 3rd millennium that is dictated by the threat of nuclear war and the mid-East conflict with Iran, Iraq and Afghanistan.

Not all the beautiful pictures that relate to each book could be included when the books were first published. Originally the pictures were taken as 35mm transparencies (color slides) with professional Canon and Nikon equipment, sometimes using a 200mm telephoto lens for close-up views. During my career from 1960 to 2000 or so, I saw the world through a camera lens, communicated by means of pictures and words, and became dedicated to recording the history in which I participated. Throughout forty-five of the fifty states and twelve foreign countries, caught up in often almost unbelievable situations, I kept snapping pictures. In a sense, then, the trilogy is a record of my own life.

Lastly, one of the major reasons I wrote the Trilogy was to leave for my children and grandchildren a sense of the struggles, fears and heartaches we faced as a nation in the 20th century. My children were all born in the last half of the 20th century and were too young to have memory of the trials and challenges their parents fought and lived through to preserve this great country in the face of sometimes overwhelming obstacles. The freedoms and privileges we enjoy today were bought and paid for with human sacrifices. This legacy is, in a sense, a birthright they must be willing to fight for and to protect in future years. This is how it was, told through pictures and words.

Tom Becker

Pottstown, Pennsylvania

2008

ACKNOWLEDGMENTS

For this third book of the Technology Trilogy, I've once again called upon commercial and government agencies for special materials and advice. Numerous internal departments at NASA and its widespread locations, especially the Jet Propulsion Laboratory at Pasadena, California furnished many kinds of photographic and informational assistance, especially Les Gaver, former Chief of the Audio Visual department of the Public Information office. Many times I was allowed to roam the launch pads at Cape Canaveral, especially during preparations for the launch of Apollo-Moon missions, Apollo Soyuz Test Project and the Viking/Mars missions. I appreciate the help of many unnamed employees at such times including pre-launch press conferences.

The National Climatic Data Center at Ashville, North Carolina was instrumental in furnishing photos and satellite data including the full set of data relating to the 1995 Hurricane Season and other seasons too numerous to mention. The credit line "NOAA/NESDIS/National Climate Data Center" refers to this organization. The Embassy of the USSR in Washington D.C. provided specific photographs and data on request and I thank them for their timely assistance. I'm grateful also to the Australian Ministry of Defence at Canberra, the U.S. Central Intelligence Agency and the Department of Defense at the Pentagon for permission to reproduce photos and specialized documents and to tour the North American Radar Defense Command in Cheyenne Mountain, Colorado as well as the U.S. Air Force Space Command Headquarters at Colorado Springs.

I'm indebted to the White House Press Corps and the Office of the President for permission to photograph selected paintings inside the White House. Equally important for me was the opportunity to meet

Tricia Nixon Cox and others during ceremonies in the Rose Garden related to my work with the U.S. Treasury Department's Bureau of the Mint and the writing, photography and printing for the *Our American Coins* publication in 1972.

Once again my debt to the European Space Agency is beyond description for continued assistance with clarification of data regarding many European space missions but especially the Giotto spacecraft to Halley's Comet. ESAs Paris and Washington DC offices sped up the flow of information from Europe and helped in the timely delivery of visual material from the Space Operations Centre at Darmstadt, Germany and close-up photographs of the Comet's nucleus from scientists at Germany's Max Planck Institute.

Specialized and national space agencies in Japan, China and India kindly responded to my several requests for information and visual materials related to unmanned spaceflight and earth observation technologies. These materials were shared with more than 200,000 students in America and Europe and helped us grasp the meaning of the concept of "the human reach into space." I owe them a debt for their many kindnesses that simply cannot be paid.

Throughout the preparation of the entire Trilogy series over the past several years, I was given immeasurable help from Barbara and Kevin Cairns, and Marilyn Becker, not only for untangling various computer problems but also for advice about a number of creative snags and other support. They should know their help is very much appreciated.

1 A HANDSHAKE IN SPACE

"If you want your scientific report for the day; Zero-G does not disturb the adult female mosquito. We thought we'd feed it ourselves here for a few days and then we'll give it to the fish. The other alternative is to bring it back alive and give it a pair of astronaut wings."

ASTP Astronaut Deke Slayton

Art rendering of the historic meeting of the Apollo spacecraft and crew (left) with the Soviet Soyuz spacecraft and crew in 1975. Painting by Bob McCall for NASA, photo courtesy of NASA.

The launch of Apollo 17 to the Moon in 1972 signaled the end of astronauts in space capsules being carried to the lunar surface. It was the end of Apollo too except for an extra capsule that had at one time been destined for launch as Apollo 18, but it never happened. The Congress, short-sighted and without a vision of the future, quickly pulled the funds out from under the NASA manned space program after the first moon landing and the astronaut corps came to an abrupt halt. As far as national leadership was concerned, America had demonstrated its superiority in space and won the "space race" without doubt. But that was the end of it! In Washington, the wheels were always turning and a nagging question hung around for months – what to do with that last Apollo craft?

Lights burned late in the Capital building. America was still in the midst of the Cold War – still in the middle of the space race – still tied up with the Vietnam War. On the home front, interest in space was slowing down at a fast rate. The cost of the war played havoc with the national budget and the American people began to worry about nuclear war again. The disappearance of NASA's budget and the last lunar mission told the world that the entire "moon thing" was finished.

President Nixon visited China in 1972 to rebuild political relations, and Soviet Premier Leonid Brezhnev came to America for a visit in 1973. By then the idea was conceived to carry out a space rendezvous between the last Apollo spacecraft and a Soviet Soyuz 19 spacecraft. It would be a tricky maneuver; both spacecraft would have to be modified to be able to connect with each other. The flight and its missions were symbolic and a fitting close to the space race that had existed for so long between the two nations. The American public, however, was not impressed with the concept of ASTP and considered it a blatant political stunt without any real value beyond an attempt at political closure with the Soviet Union. The technology would be interesting and, in the event an emergency in space warranted immediate rescue, it would be good to have a usable plan. Americans wanted another Moon mission; the Congress wanted to pay for the Vietnam War.

In the engineering books, the rendezvous is described as ASTP – Apollo Soyuz Test Project - a test of the multiple docking module that would connect the two craft. For the first time, the Soviets agreed to parade publicly what they had been hiding all those years – Soviet space sites and technologies. In an early gesture, American astronauts Donald "Deke" Slayton, Vance Brand and Thomas Stafford began studying Russian language while Soviet cosmonauts Valery Kubasov and Alexei Leonov began learning "American" language, which is somewhat different from learning English. By the fall of 1972 they all were deep into learning their new languages although the crews weren't publicly announced until May 1973.

In order for the two spacecraft to complete a solid docking, a new docking module was designed in which one end snugly fit the Apollo craft and the other end safely fit the Soyuz spacecraft. Computer data and laser beams were set to record the distance between the craft and announce when the two craft were perfectly aligned just prior to docking. In practice, the alignment was so perfect that the Apollo craft was able to simulate a temporary eclipse of the Sun so that the Soviet crew could study the Sun's corona.

According to the mission profile, the astronaut/cosmonaut teams were scheduled to complete 27 separate experiments during the mission, of which five of them were conducted jointly. Both teams visited each nation's space facilities. The Soviet crew spent time at NASA's Mission Control Center at Houston, Texas, and the NASA Launch Control Center at Cape Canaveral (including the Vehicle Assembly Building and launch pad). The American astronauts visited the Soviet Mission Control in Moscow, cosmonaut training center at Star City just outside Moscow, and the Soviet launch center at the Baikonur Cosmodrome (Tyuratam) in Kazakstan.

With all these activities in mind, the USSR Academy of Sciences and the NASA signed a mutual science research and low Earth orbit rendezvous/docking agreement on May 24, 1972 although crews were not announced until 1973. Goals of the test project were 1) to test the systems compatibility of the two spacecraft for future rendezvous and docking, 2) design and construction of a compatible international docking module to fit both spacecraft, and 3) development of additional in-flight techniques, communication systems, and overcoming problems of language. For example, during the mission the language rule was: use the language of the person you are talking to. The policy was highly successful.

Both craft were launched within 7½ hours of each other on July 15 1975 (an American nod to 1775). A three-stage rocket lofted the Soyuz 19 craft; the U.S. Apollo craft launched by a Saturn rocket carried the new docking module fitted with hatches at both ends. In this design, the docking module served two purposes: 1) to fasten the two craft together during the mission, and 2) to provide a passageway (tunnel) between the two craft so the two crews could visit each others' spacecraft. During the mission, both crews took advantage of this provision when the meeting in space took place on July 18.

As the mission progressed, astronauts and cosmonauts floated back and forth to visit each other's spacecraft. At one point, astronaut Stafford and cosmonaut Leonov met in the open hatchway of the docking module and shook hands while television cameras recorded the momentous event. Plaques and flags were exchanged along with other special mementos including an exchange of boxes of tree seeds, and the crews ate specially prepared meals. The Soviet Soyuz spacecraft landed after six days about 300 miles east of Baikonur Cosmodrome on July 21. The Apollo spacecraft splashed down about 330 miles west of Hawaii on July 24; the crew was picked up by the USS New Orleans. Both crews eventually were alive and well, but just barely.

The Apollo crew spent a total of nine days in orbit finishing up several experiments. A severe problem occurred for them during the landing, however. At about 23,000ft the Apollo craft began to leak fumes described by the astronauts as "a yellow gas" that irritated their eyes and made them choke and cough. Already well into the splashdown phase of the landing and floating down on parachutes, the Apollo craft began to fill up with fumes. To make matters worse, the spacecraft somehow hit the water and rolled over upside-down with its heat shield pointing skyward and the exit hatch under water. The crew's normal route to get out of the spacecraft was closed off.

The crew of Apollo 18 was in a highly dangerous situation unless they could roll the spacecraft over using special flotation balloons. Strapped in their seats, the men were unable to reach their oxygen masks. Stafford quickly unstrapped himself and fell to the floor grabbing the masks on the way down. He gave one mask to Deke Slayton and shared the other one with Brand who had passed out. As Brand regained consciousness, the Apollo craft righted itself and the crew was picked up but it was a tense several minutes nevertheless.

Word reached me in July 1972, before the crews were announced publicly, that the mission was being organized in both countries. Sometime later, an agreement was reached between me, master sculptor Edward R. Grove of West Palm Beach, Florida and officials at Presidential Art Medals of Vandalia, Ohio to produce an art medallion to commemorate the unusual international space mission. We had precious little time to prepare the designs and work started immediately. Ed Grove would design the obverse (front side) and I would design the reverse side.

My part took a great deal of thought, but I finally decided to create the obvious – an in-flight orbital view of the rendezvous of the two spacecraft. All well and good, but what about the background? Research began in earnest even before NASA printed the mission profiles. What was the medallion going to be telling the world? That too was obvious – two nations at serious odds with one another still could put aside their differences in order to cooperate on a scientific and cultural level. At the same time, the ASTP mission would be proof to other countries at odds with each other that there always was hope that differences could be overcome. Immediately the Middle East came to mind, not only as a place on the planet but also almost a contradiction to peaceful endeavors. The location would have to be biblical – a place that epitomized human strife for centuries. I chose Arabia and the Sinai Peninsula as symbols of humankind's ages-old battlefields of global discontent.

**The oil painting was an attempt to give the sculptor a
visual concept of the reverse side of the medal.**

Over the years, I had visited with a number of highly
respected coin and medal sculptors – Chief Engraver Frank
Gasparro of the Philadelphia Mint who designed the Eisenhower
dollar coin, Ralph Menconi of New York who sculpted numerous
medallic series including the entire set of Apollo medallions,
Felix Schlag who designed the Jefferson nickel, Paul Vincze
of Chelsea, London, former Chief Engraver Gilroy Roberts
also of the Philadelphia Mint, and others. Their coaching and
commentaries helped a great deal to shape my thinking. I also
corresponded with Chesley Bonestell, the dean of space art, and
studied his artwork at the National Air and Space Museum. It
was the encouragement of these giants that allowed me to grasp
some of the fundamentals of coin and medallic art. Putting
their advice into practice, however, was a huge challenge. I had
already designed a number of medallions but was struggling with
the technique of oil painting.

Before working on the ASTP design, I had designed both
sides of a set of three medals commemorating the three missions
of Skylab in 1972/73, America's first station in space. My part
of the ASTP design began with an oil painting that eventually

became the final design. Recalling Bonestell's perspective, I went to work and at last completed the painting. Now the painting image had to be sculpted in plaster in order to begin the process that leads to the stamping dies. Pictures of the painting were sent to Ed Grove in West Palm Beach, Florida and the rest is history.

At the Cape, the launch pad "rollback" display gave us a good view of the lone Saturn rocket bathed in the cool but harsh glow of the pad searchlights Dozens of photographers set up their tripods, checked lens settings and shutter speeds, and settled down to do their work. In front of us stood the brilliant display of the enormous Service Gantry surrounding the Apollo 18 spacecraft and its supporting Saturn rocket. Ever so slowly, the giant gantry moved backward leaving only the Apollo Saturn standing alone on its support framework.

As the launch date came closer, NASA cleared me for admittance to the Cape as a photojournalist and I prepared for what became my last working visit to the Kennedy Space Center. Arriving in Florida several weeks later, I realized that the weather was going to be very hot – much hotter than expected. As part of the Press Corps, we were awakened once more at 5:00am to take a bus out across the pond from Launch Pad 39 for a photography session.

For the trip to the Moon, a three-stage rocket would support the spacecraft but for just a trip to low Earth orbit only two rocket stages are necessary. The magnificent sight was once again "otherworldly", set as it was against the pale morning sky with the first faint rays of a dawning day behind it (see the front cover photo). Occasionally a lone sea gull passed between us and the launch pad, hinting that there was life there in the midst of all the engineering and the chemicals that made up the rocket fuel. It seemed a view no artist could possibly capture on canvas.

As the day wore on, we were transported back to the Press viewing stand to set up our cameras again and await the launch countdown. The near-cloudless day promised to be steamy hot and by noon the Press Corps group was suffering in the forbidding heat of the day. Most of us sought shelter beneath the viewing stands to escape the sun; some of us were beginning to feel light-headed and nauseated signaling the first stage of heat exhaustion. At launch time, nearly 4:00pm in the afternoon, the day's activities began to wear on us and we barely had presence of mind to carry out the tasks required to complete our photography.

From three miles, we watched in awe as fire and steam erupted around the launch pad and Apollo 18 slowly lifted into the blue skies, speeding onward toward its historic rendezvous with the Russian Soyuz craft itself already heading for low Earth orbit. The stinging crackling of the Saturn engines once more bore into my head and pinged off the massive Vehicle Assembly Building feeling much like a dentist's drill and forcing me to cover my ears. I could feel the concussion from the rocket's blast as it hit my clothing, and almost as suddenly it thankfully subsided.

The duties now over, we headed for the buses waiting to carry away the many members of the Press Corps, mindful that we were the last Apollo eyewitnesses to an historic event. ASTP was my fourth and last Apollo launch experience. Flying back home, I couldn't help remembering those four launches I had attended of men going in harm's way to assure America's (and humankind's) journey toward a tomorrow in space that has no end. ASTP was a technological triumph but a political and cultural failure. No stern walls of disagreement tumbled down and there was no easing of tensions between the two superpowers despite everyone's efforts.

Our human and mechanical assault on the endless expanse of space was only the barest beginning of an adventure that held great promise – and great peril – for the future of humankind. What would we do, I wondered, with this tiny beginning that offered such a wealth of adventures to come? Six pairs of astronauts walked across the chalky surface of the

Moon, amazing new technologies were created and the people of Earth had their first glimpse of starry heavens, but the fact still remained we could not get along with each other here on Earth. If we could only survive our own emotions, if we could turn away from the violence of the battlefield that has plagued us down through the centuries, we just might someday be able to visit other worlds and discover what we might even think of as the human destiny. We are always capable of the most brilliant of creative technologies, and the most blackened of human atrocities that perhaps co-existence is merely an illusion after all. Still, it is possible. . .if only. . .

Left, (reverse), the Apollo 18 and Soyuz 19 spacecraft meet in low Earth orbit over the Sinai Peninsula. Text means the same in English and Russian; design photos by Tom Becker.

Right, front side (obverse) of the ASTP art medal, design and sculpture by Edward R. Grove. Robert Goddard, America's space pioneer left and Konstantin Tsiolkovsky, the Soviet Union's space pioneer right, are featured against a symbolic background of mutual agreement.

Celebrities and government officials usually attended the launches, more for orientation than just for the sake of appearance. National Security Advisor Henry Kissinger meets with the press in this photo taken at the launch of Apollo 14 in 1971.

2   RENDEZVOUS AT 20,000 FEET

"While you're at it. . .can you check the tires and clean the windshield?"

Pilot of a B52 Bomber

A fully armed B52 bomber approaches the boom nozzle of the
KC-135 refueling tanker at about 20,000 feet altitude.

In the middle of summer 1978, it was a beautiful day in Grand Forks, North Dakota which was rare enough even for summer time. Today, however, was an extra special day and I was thankful for a clear blue sky with unlimited visibility. I was invited by the Strategic Air Command to do some research, participating in a series of airborne rendezvous and refueling missions - in fact, several missions in one. It was going to be a grand adventure I looked forward to with growing anticipation.

Airborne refueling was first used in World War I. Two aircraft met at a point in the sky and one passed fuel to the other through a long hose. Although the concept isn't new, the particular technology has advanced right along with other newer inventions in aviation in general. Instead of transferring fuel at 10,000 feet altitude the exercise today is done at a much higher altitude – say, between 20,000 and 23,000 feet just below the level where crews need to go on oxygen. The only real difference today is the danger level because jet aircraft are moving much faster and carrying many times the amount of fuel.

When someone shook my shoulder at 5:30am, the call to duty was a sudden reminder that "the grind" was about to begin. Breakfast in the mess hall at 6:00am (scrambled eggs, bacon, toast) – in-flight briefing room at 6:30 – on the flight line at 7:00 with notebook and cameras. It was a scramble but we all made it on time. The Boom operator also worked as a navigation confirmation, shooting the Sun with a sextant and checking with the aircraft navigator-engineer's radar. By this procedure, the two men continually reconfirm the position of the airplane not just in the sky but more importantly over the ground. It's also a quick check with the flight engineer to make sure on-board equipment and electronic calculations are correct.

The crew consisted of pilot, co-pilot, flight engineer, boom operator, and eyewitness cameraman. All were aboard. I was surprised they all were almost half my age. A SAC officer reminded me that most of the planes we were going to refuel were much older than the crew members flying them. Engines started at 7:10 and we moved out along the taxi strip heading for the runway. The plane was a KC-135 tanker, the equivalent of a Boeing 707 with 100,000 gallons of high octane aviation fuel on board instead of passengers.

By 7:30am we were airborne on the flight leg to our first checkpoint and the workday was just getting started. For the crew, this would be just another practice mission but highly important to keep the crews trained. In time of war, it is possible to keep an airplane in flight for several days without landing simply by a series of airborne refueling maneuvers. The refueling capability allows aircraft to reach more distant targets, perhaps making more than one bombing run on the same or adjoining targets. For airborne reconnaissance, the maneuver is highly important because it permits the recon aircraft to penetrate deeper into enemy territory or even to photograph several different targets in succession that might be many miles apart.

Our overall course was generally over the Midwest – to Des Moines, Denver, Chicago, Grand Forks, and out again. We saw a lot sky that day. I realized I was inside a flying gas station at about 20,000 feet, cavorting across the United States at better than 400 knots ground speed, ready to meet two other aircraft flying at the same speed and loaded with a formidable arsenal of weapons and sensitive electronic equipment. All these men were well trained, and were some of the most competent individuals I had ever seen anywhere. They went to work every day like the rest of us except they happen to work in the wild blue yonder. It was reassuring to know they were so capable; heaven forbid if worse came to worse, however, they would do what had to be done and do it well.

The call sign for the refueling tanker was "Patch-Two-Two" and it didn't take very long to hear it over the earphones. "Patch-Two-Two this is Bronco Four – I have you visual. We'll meet you going into the ARCP (airborne refueling contact point)." Bronco Four was an E-4 Airborne Warning and Control System aircraft (AWACS) coming up fast on our tail and several miles out. We had already picked him up on radar sometime before he called us and were aware of his long-range presence.

"Roger, Bronco Four, we have you visual now. Holding course and speed," the tanker pilot responded. I had crawled to the rear of the plane wearing a crash helmet, a 28-pound parachute, earphones, two cameras and a notebook. Beside me was a Tech Sergeant who operated the long boom carrying the gas hose and nozzle. We lay on our stomachs side by side looking out a glass window that gave an unrestricted view to the rear through the specially designed window. "Boom", as the crew called him, held a steering handle like the joystick for a computer game (look back at the lead photo for this chapter) and by which he guided the large boom.

The boom had two guide vanes; one on each side of the "pole" that its operator could use to point the boom in any direction making it an aerodynamic structure allowing the operator to "fly" it with the joystick. The actual hose that carries the fuel can slide back and forth inside the pole. Once the receiving aircraft is lined up and at the proper altitude, the boom operator can quickly guide the hose nozzle toward the other airplane. The maneuver sounds simple but at these speeds, and with the receiving craft flying directly in the tanker's slipstream, things can sometimes get pretty dicey. In this kind of close quarter maneuver, there is no such thing as "…that's close enough"; everything has to be precise.

The AWACS aircraft was about the same size as a Boeing 707 fitted with a huge horizontal disc on the outside roof that continually spun around like a frisbee. On top of the AWACS cabin over the pilot's seat is a spring-loaded trap door that opens up to receive the hose nozzle.

"Pilot to Boom," the tanker pilot radioed, "bring him on in – you're flying the airplane now." The Boom operator countered with "Roger – I'm flying the plane; holding course and speed – Bronco Four is inbound at a thousand feet." The pale gray AWACS carefully closed the distance between the two aircraft as the Boom operator kept a running chatter to guide the oncoming airplane. Finally the AWACS loomed large outside the window seeming to almost break through the glass while the Boom operator gently coaxed the other pilot to a line just below our tail.

"Two hundred feet, up two. . .one-fifty, up fifteen. . .one hundred, up five. . .sixty feet, steady. . .thirty feet, up two. . .twenty feet, up two. . .ten feet, steady. . .contact, contact, hook." The fuel nozzle slid out through the end of the boom into a trap door that tightly clutched the nozzle. "Roger Two-Two, contact," the AWACS pilot confirmed. "Transfer on," the boom operator announced as the two aircraft gently hooked together.

"While you're at it. . .can you check the tires and clean the windshield?" the AWACS pilot bantered. Not to be outdone, the

Boom operator countered, "Looks like your left rear tire is low. . . how about getting out and opening the trunk so I can get to the spare?"

In just under five minutes, 20,000 pounds of high octane fuel was transferred from one aircraft to another. "Bronco Four, this is Patch Two-Two - numbers follow," the boom operator ended the silence. "Twenty spike zero, zero, zero. . .time fifteen thirteen dash fifteen seventeen point five. . .standby for release on my mark."

"Roger Two-Two."

"Standby. . .mark, mark, release."

"Thanks a bunch, guys. . .see ya 'round the neighborhood."

Very slowly, the AWAC glided downward and turned off to the left. There was no noticeable jerk or movement when the disconnect was made. Boom retracted the fueling boom and tucked it away under the tanker's tail. The operation was over and Bronco Four was on its way to another part of the sky and another mission.

Lunchtime in a box – three sandwiches, chocolate and white milk, tossed salad, hard-boiled egg, an orange, cup of butterscotch pudding, and a "flight menu questionnaire." The lunch cost me $1.05 as a civilian, the crew members' was free. Then off to a new rendezvous in another part of the sky and a repeat of the same maneuver all over again, this time with one of America's most feared weapons.

A B52 bomber is huge. On the ground, I sat in the cramped pilot's seat and had a chance to study the instrument panel, overhead dials and switches as well as the radar screen, forward-looking television image and other controls next to the pilot's seat. The complexity of the pilot's cabin was awesome considering the many instruments the pilot and co-pilot had to monitor and respond to.

Now we were looking for a fully armed bomber that soon appeared as a tiny olive drab and camouflaged speck below and far off to the left, barely visible as mostly a dark green against the occasional clouds and background of the Earth. It was growing steadily larger and sliding straight for us.

The same rendezvous and connection routine was followed once again. I studied the B52 bomber as it closed in, fitted with rockets on the wings and loaded with high explosives in the bomb bay area. Without a doubt it was a mean machine. After the Vietnam War, it was disclosed that the one thing the North Vietnamese soldiers feared the most was the B52 bomber. It could drop conventional bombs as well as napalm bombs and defoliation incendiaries, carrying a heavy load intended to tear up a large amount of real estate.

At the end of the fuel transfer, the bomber disconnected and it, too, slid off and downward to the left. By now it was late in the day. The crew had successfully completed its several navigation and rendezvous missions and was turning toward home at Grand Forks. It was just as well since I had run out of film and was tiring from the intensity of the day's activities. Now we were "heading for the barn" and tomorrow was another day of continual challenges, new technologies, new electronic systems and. . .a new kind of uneasiness I hadn't anticipated.

One of the clearest indications of disagreement among nations at the height of the Cold War was the fact that the United States and the Soviet Union had thousands of nuclear missiles pointed directly at each other, all ready and able to launch at a moment's notice. If let loose, the missiles could depopulate about 40% of the Earth's surface in a matter of a few hours. In both countries, underground missile sites were spread across the countryside with great forethought by the military-industrial complex. Early the next morning we were on our way to visit an operational missile site at Minot, North Dakota outside Grand Forks.

After driving what seemed a reasonably long time, we turned off the main highway onto an innocuous dirt road leading up to a chain-link fence. The sign attached to the gate warned of a government installation. Inside the fenced area was a single low green-painted building looking for all the world like someone's vacation cottage. As we approached the fence a uniformed guard with an automatic weapon appeared as if out of nowhere along with a guard dog visibly upset at us for being there. After showing our ID passes, the guard reluctantly opened the gate and we were taken inside the little building.

There was an elevator to take a Captain and a Lieutenant, both carrying sidearms, and me down eight stories underground to the *missile launch control center.* As we were descending the eight stories, I noticed both officers unbuttoned the holsters holding their .45 caliber pistols. I asked why they did that. "If you try to reach for the launch console, we'll shoot you," he answered with a slight smile. I wasn't especially amused.

Arriving at ground level, we went through a doorway into the launch control center where two officers each sat in front of a maze of instruments, dials and buttons. One console was devoted to communications, the other console had 20 sets of levers – one lever for each remote missile silo meaning the launch operator could personally launch 20 separate pre-aimed missiles. Both consoles had a key slot where a key could be inserted and turned to arm the missile launch system. The consoles were sufficiently far apart so that no one person could turn both keys at the same time and personally arm all the missiles by himself.

The two officers went through a well-rehearsed briefing explaining how the launch system worked, pointing to the various dials and switches in turn and ending with the launch keyholes. This particular site was in charge of 20 Minuteman III ICBMs (Intercontinental Ballistic Missiles), all of which could be launched within a matter of a few minutes and all pre-set to follow a precise computerized pathway to its target destination that could be as much as 6,000 miles distant. Most likely that target was directly on the other side of the planet.

As we finished up the briefing, we stepped out into the hallway where an already-opened door revealed the naked missile sitting in the launch silo waiting for the moment it was called upon to fulfill its function. A small walkway big enough for only one person led from the hall straight over to the missile. Fascinated by the presence of an actual nuclear missile, I crossed over the walkway (called "the diving board") and put my hand up against the side of the missile warhead.

The missile was humming ever so slightly with the heartbeat of its electronics and I wondered how it would feel as it sped in flight toward its target. Leaning over, I could look down the full 60 feet length of the missile into the depths of the silo. An involuntary chill went through my body – this scenario was for real – a live missile carrying a nuclear warhead capable of killing thousands of people in the blinding flash of an instant. The warheads are contained in several Multiple Independent Re-entry Vehicles (MIRV) which shower down on the target and detonate independently. The experience was at once reassuring but humbling, especially when the thought crossed my mind that the Soviets had about the same number of nuclear missiles aimed at the United States.

Headquarters for the Strategic Air Command is at Omaha, Nebraska where most of its real work is carried on eight stories below ground in the SAC *Underground Command Post*. From there, SAC can reach anywhere in the world with a single phone call or radio communication. This ability includes aircraft in flight, ground stations, foreign allied government strategic facilities and even foreign aircraft in flight. SAC also can reach out and touch other intelligence units such as the National Security Agency, Central Intelligence Agency, Britain's MI-6 group, or the Israeli Masad.

SAC underground Command Post at Omaha, Nebraska features data screens 18 feet high on which can be projected strategic information for making global battle decisions. Photo courtesy Strategic Air Command, U.S. Air Force.

Centerpiece for the command post is the set of gigantic eighteen-feet tall projection screens on which data is shown detailing target environments, attack strategies, reconnaissance imagery, and bombing raid monitoring. Analysts and technicians sit in front of the screens and call up documents, photographs, tracking data, or bomb runs. Telephone and radio signals right from the battlefield can be projected for analysis and development of special engagements with enemy fighters.

During the previous year (1976), a group of specialists was flown to Omaha for a briefing on SAC capabilities and the Soviet threat with an analysis of Soviet weapons and space policy at the height of the Cold War. I was privileged to be in that group and spent three days of intense indoctrination, especially about the Soviet threat and capabilities. We viewed an amazing array of technologies and weapons arsenals on both sides of the Iron Curtain as well as a comprehensive tour of the SAC facilities in Omaha.

General Curtis LeMay coined the phrase "Peace Is Our Profession" for the Strategic Air Command back in the early 1960s. It was an excellent description of the mission of the SAC teams. With the kinds of assets and deterrents available to the SAC community all over the globe, foreign countries have been reluctant to tangle with the kinds of weapons we are able to view on television during the Iraq wars. Considering the readiness of SAC bombers and fighter planes that can be anywhere on the globe at a moment's notice, peace always seems a wiser choice for foreign nationals contemplating armed intrusion.

As I left the several days spent at Grand Forks and Minot, an ominous feeling of foreboding hung over me that followed me for days like a darkened cloud. The men I met were exemplary – competent, aware, well-trained and highly knowledgeable not only about their missions but about global physical and political conditions as well. They wore their many responsibilities like a lightweight burden of duty. The seriousness of their daily work demanded an unusual accountability and maturity for their young ages, but their responsibilities were not a heavy-weight burden. It was reassuring to watch them work with such ease and correctness of purpose.

The arsenals were unnerving and numbed me into silence as I thought about their potential destructive powers. In the 1970s, the world was divided into two armed camps, each camp with the ability to destabilize the global situation and cause the kinds of havoc for which there is no precedent. I felt highly protected and have to admit that it was comforting and conducive to a good night's sleep to know that such men as these were willing to put their lives on the line so the rest of us can lead reasonably productive lives out of harms' way.

On occasion I've been known to comment, "What I carry around in my head every day would keep you from a good night's sleep." The very thin edge of the balance of power in the world is the narrowest of lines between life and complete annihilation. There is nothing in between these two choices. It is only because of people of maturity and good will that no one will ever have to make that choice.

People living in the freedom of the free world have little or no concept of the extent of the constant threat from the outside world. The policies of our intelligence community stipulate, for the most part, near-complete silence about conditions of global threat and the part played by individual agents to keep abreast of the wolf at our door. We never hear about the successes of the CIA, NSA or the DIA, much less the competence of the support from their vast and complicated infrastructure. So we have no measure of the intelligence community's successes. Let it suffice to say that America is well guarded and protected and, as Robert Kennedy put it so simply, "America is in good hands."

The American population is cut out of the communication loop of the intelligence community because of a "need to know" policy. Personally I know enough about the workings and missions of the intelligence community not to want or need to know the rest. For example, how many nurturing plans to attack homeland America have been foiled by alert and hard working intelligence operatives? How many operatives have lost their lives in the covert defense of America? How well guarded/protected are our ports, military bases, bridges, nuclear plants or water purification systems? I have no idea – but I know they ARE protected and I don't need to know the details to feel safe. As this book is being written, there are dozens of defense/attack aircraft (like the AWACs and B52s described in this book) plying the skies over American cities and key installations *every hour of every day*. Just look up and you can see the contrails criss-crossing occasionally in the skies, made by military aircraft practicing rendezvous or simulated bombing runs. They allow me a good night's sleep, and I am very thankful for that!

The teeth of the dragon, looking straight down into the missile silo.

3 1600 PENNSYLVANIA AVENUE

"Just as the oceans opened up a new world for clipper ships and Yankee traders, space holds enormous potential for commerce today."

President Ronald Reagan,
Address To Joint Session of
Congress, January 25, 1984

President Reagan presents his vision at the White House about
"America's Future In Space in 1984." Left, NASA Administrator
James Beggs; right, DOT Secretary Elizabeth Dole looks on.

I've never been especially impressed with government, ours or theirs, or the day-to-day mechanics of government. Perhaps it's because we always expect much more from politicians than they're able to give. Government employees are just people like the rest of us but the power

23

they have at their disposal can be awesome. At other times, there is a certain kind of magic surrounding the workings of government – it has a "get things done" kind of aura that leads us to believe it can solve any problem quickly, which often is a mistaken belief. Occasionally the wheels of government grind exceedingly slow. Also we seem to hold Congressmen and Executive Branch incumbents in more than ordinary high esteem which can be a mistaken attitude.

On the other hand, the design and architecture of Washington DC is a most impressive and imposing sight and gives one a feeling of truly being in Camelot. The memorial buildings and their locations are reminders of America's heritage along with all the national museums and government buildings. The city contains America's entire history. It is nearly impossible to read the inscription on the wall beside the huge statue of the seated Abraham Lincoln in the Lincoln Memorial without feeling one's throat tighten with emotion. The same can be said of the statue of Thomas Jefferson, standing alone in the center of the Jefferson Memorial. Washington DC is a showplace of who we are as a nation and a people, and the ideals we have died for over the past several centuries.

Visits to Washington DC in the 1970s and 1980s were always filled with the personal excitement of new ideas and new projects and the promise of great adventures. In the forty or fifty trips to the city, the experiences I had there were always life-changing, not just because of the wonder of the surroundings but more because of the work I was involved in.

The Director of the Bureau of the Mint in Washington DC in 1971 contacted me to create a booklet reviewing current U.S. coins. After engaging a local professional artist, and after several creative meetings, we arrived at a workable concept involving several photographic sessions in Washington and at home and my writing the full text. We signed the contract with the U.S. Treasury Department and work began immediately on several days' photography of presidential monuments. It was a challenge considering the weather, light sources and waiting for tourists to vacate the monuments. On one occasion, however, the Treasury building was blocked when I arrived because of a bomb threat and I had to undergo an electronic body search and close inspection of my briefcase.

We needed pictures of oil paintings hanging in the White House for illustrations; the task would involve getting permission from President Richard Nixon to roam the White House with cameras and a ladder. The arrangements were made and I soon found myself checking through the front gate and entering the building where I met a Secret Service agent and a very kind maintenance employee with a tall ladder and we started our tour. With a list of the paintings, we simply visited each in turn from the top of the ladder and the job was finished. The experience was almost mind-boggling; it was like having permission to roam The Louvre at will without any restrictions or hindrance. Truly – it was Camelot. Oddly, I never took an official tour of the White House because I never had the time to go to Washington as a tourist.

Huge likeness of Lincoln adorns the Lincoln Monument in the nation's capital.

Each monument was photographed in turn giving me an opportunity to read the sculptured texts in detail in addition to studying the statues themselves. Previously I had been teaching American History in high school, but our nation's history now took on a far deeper and greater meaning. The things these men had done, and the words they uttered and wrote - often in trying circumstances – were bigger than life itself. By the time I was finished touring the monuments, I was reminded once again what it meant to be an American with such a powerful heritage. "Humility" is a word that doesn't even come close to describing my feelings.

Over the following months, the text, photos and design all came together for the booklet that was published in 1972. To celebrate the official public release of the booklet, a small ceremony was held in the Rose Garden and once again I found myself in Camelot. The Director of the Bureau of the Mint Mary Brooks and Assistant Director Roy Cahoun, various Treasury Department employees, and President Nixon's daughter Tricia Nixon Cox were all in attendance on a warm, cloudless day that now seems much like a dream. The title of the booklet was printed as *Our American Coins*. Bureau of the Mint policy at the time required that authors' and designers' names not be included on government publications.

On occasion during business trips to Washington there was barely sufficient time to roam the city and its buildings. The visits often took me to such locations as the National Air and Space Museum (Old and New), the Capitol Building, U.S. Government Printing Office, U.S. Treasury Department, and numerous history museums, always with a camera at my side. There is no other place in the world like Camelot but I seldom had the opportunity to visit these places as a tourist. I vowed during each visit to come back some day as just a tourist but so far it still has not happened.

In the fall of 1984, a letter from NASA arrived in the mail in the form of an invitation to attend a major Presidential presentation with the title "America's Future In Space", to be delivered by President Ronald Reagan at the White House. Other dignitaries also would be present: Department of Transportation Secretary Elizabeth Dole, NASA Administrator James Beggs, and the Assistant Secretary of the

Department of Commerce. The presentation came at just the right time as a numbing public apathy about space technologies had set in despite the introduction of the new Space Shuttle Orbiter. On the other hand, American intelligence sources indicated a crash Soviet effort was in progress to complete the Russian Space Triad: a manned space station, a Shuttle-type transport vehicle, and a heavy lift launch rocket. Intel reports were never mentioned in the press or during the President's speech but our intelligence community was hard at work in liaison with underground Russian nationals.

President Reagan had already announced America's *Strategic Defense Initiative* (SDI) in March 1983 as a land- and space-based protective nuclear missile system to be instituted as a guard against foreign missile intrusion that might threaten U.S. national security. The "Star Wars" shield as the media referred to it would actually be a missile and satellite umbrella capable of thwarting an incoming missile attack. SDI was designed to replace the outmoded *Mutually Assured Destruction* (MAD) philosophy which the President referred to as a "mutual suicide pact." The SDI design would have as its base a sophisticated laser system to shoot down incoming foreign ICBMs before they could deliver their destruction. Although the Strategic Defense Initiative was never actually implemented, it was a point blank message to the Soviets that we were getting prepared for such an attack and would allow America to gain its continuing programs to conquer the high ground. The Soviets already had established a similar system to protect Moscow back in the late 1970s.

With a camera and a tape recorder, I arrived at the White House along with a large contingent of other photojournalists and television broadcasters. As we reached the check-in security station, we were asked to operate cameras and tape recorders to assure the security force that the equipment was genuine. At the time the measure seemed a little extreme, only later to find myself seated a short ten feet (center) from the President while he delivered his presentation.

President Reagan's stature was almost overwhelming. He was a huge man with broad shoulders and standing more then six feet tall he looked all the more like a football quarterback. Very much a soft-spoken man, his calm voice belied the sheer determination and dedication carried by his words. As he worked through his speech, a new frontier in space opened up before us, including a manned space station, the emerging era of the Space Shuttle, continued solar system research, and a strong program to send spacecraft out into the endless cosmos. It was exhilarating to envision, as he said, a new frontier waiting to be explored and a new kind of space-based commerce.

He also outlined in great detail America's plans for national defense put in place by a global system of SDI – an amazing umbrella of offensive and defensive satellites with nuclear weapons to confront aggression in space from almost anywhere on the planet. This kind of an awesome overhead shield was to be America's major space-based deterrent against attack from any sudden threat at any time of the day or night.

NASA Administrator James Beggs reaffirmed the space agency's preparedness to offer full support as did Department of Transportation Secretary Elizabeth Dole, reviewing the blossoming Space Shuttle Orbiter program as an in-orbit research and transport vehicle linking the orbiting Space Station with international research organizations on Earth. The ultimate question on everybody's mind, of course, was money. Given the Congress' historic record of funding space technology projects, NASA's access to real money didn't seem especially forthcoming. Someone had to do some hard selling to get the legislative branch to open its coin purse. If everyone in the audience that day planted just two inspiring articles in major newspapers and magazines, perhaps the Congress would get the message. On the other hand. . .

4 BACK TO MARS: THE MARINER 9 AND VIKING MISSIONS

"Fact has proved to be nearly as bizarre as fiction. Although the canals were an illusion, and the possibility of life now seems less likely, the planet retains its fascination. Mars is a geologist's paradise."

Viking Orbiter Views Of Mars
(NASA SP-441) 1980, USGPO

Self-portrait on Mars taken by the Viking 1 Lander at Planitia Utopia gives a hint of the distribution of rocks on the dry soil at this landing site. At lower left is part of the meteorological instrument; at right is the scoop arm that dug into the Martian soil. Photo courtesy of NASA/JPL.

Since its first appearance on Earth many thousands of years ago, the upright-walking human species has been plagued by two very deep and basic philosophical questions that are continually asked over and over again. Where is God, and are we alone in the universe? Following these questions, a long litany of associated questions has sprung up that are not possible to answer either. Every time we discover a new planetary system out in the great expanse of space, the same questions arise over and over.

Can this new planetary system support life? Why are we alone in the universe? Is Earth the outpost of another greater civilization and, if so, where is the Motherland? What does the Bible say about our loneliness? Is it necessary for Earthlings to start up another civilization on another planet to stop our loneliness? Why is Earth the only planet in our Solar System to harbor life? What will happen and what will it mean if we DO discover life on some distant planet? Deep – very deep!

The number of very costly missions to Mars attests to the fact that human interest in the planet has never wavered and continues to hold a place of enduring fascination. Three previous unmanned flyby missions in 1965 and 1967 (Mariners 4, 6, and 7) already told us that Mars was a planet like no other in the Solar System. Everything being learned led from one amazing clue to the next as the missions slowly pulled back the curtain hiding an amazing array of Martian mysteries.

The Mariner 9 mission was launched from the Cape on May 30, 1971 on a 248 million mile rendezvous with the red planet. Cut down to basics, its mission was to 1) map 70 percent of the Martian surface, and 2) conduct a study of characteristics and monitor changes in the Martian atmosphere and on the planet's surface. A nearly flawless mission, Mariner 9 made several highly significant discoveries – a vast canyon more than 3000 miles long scientists named *Mariner Valley (Valles Marineris)* in honor of the Mariner 9 craft and that dwarfs anything similar on Earth, an enormous volcano some 600 miles in diameter – *Olympus Mons*, and the first close-up images of Mars' two moons (*Phobos and Deimos*) ever taken. But the sophisticated technology aboard Viking permitted such surprising data there was much more that the craft was capable of accomplishing.

As Mariner 9 first encountered Mars, its television cameras were turned on and ordered to begin taking pictures. Nothing – a whitish covering obliterated the planet. In a panic, scientists and engineers feverishly checked the spacecrafts' instruments and found them functioning normally. By the end of September, scientists realized that the largest dust storm ever recorded had moved across Mars traveling at 30 miles per hour. Not a single Martian feature was visible for two months. Then on November 8 a small dark spot emerged from the pall of dust that scientists decided must be *Olympus Mons*, Mars' tallest volcano. This feature was followed by three more visible volcanoes – *Ascraeus Mons, Pavonis Mons,* and *Arsia Mons* – three more mountains. Despite their great height, the volcanoes barely protruded above the storm clouds.

Scientists realized that much of what they had learned from the Mariners 4, 6 and 7 missions was incomplete and had to be abandoned. On other areas of Mars appeared a puzzle of enormous size that scientists didn't even know existed. By January 1972, the storm was clearing and the real work began. Mariner 9 became a satellite of Mars for 516 days until it ran out of attitude control gas and tumbled out of control in October 1972. In the meantime, the spacecraft revealed a fantastic world of incredible proportions. It would be the responsibility of a new mission to explore these proportions. We would have to get accustomed to finding surprises on Mars; every time we go there we find something new.

On one of my visits to NASA Headquarters in Washington DC one day, planetary geologist Dr. Thomas Young expressed what he thought we had discovered on Mars. *"We found a new world,"* he replied. *"It's a fascinating place of things we never knew existed; but it will be a long time before we make Mars give up its secrets."* Young's description and prophecy later proved truer than he could have imagined. No matter how many robotic missions to Mars are sent in the name of pure science, each mission is a completely new introduction to a planet of endless surprises. Poised on the doorstep of a completely different kind of mission, extraordinarily complex and ambitious, scientists were led toward new horizons in technical engineering and in the geological and biological sciences. The new mission was called Viking.

Up to this point, four Mariner missions confirmed the basic truth of Mars' surface; impact cratering was one of the major forces that sculpted the planet into what it is today. Our coverage wasn't broad enough, though, and the expanse of territory not sufficient to provide a greater view of the planet overall. We needed to go back to Mars again to answer the same old questions that haunted us from the very beginning:

> -is there evidence of life on Mars, at any stage of existence?
> -how did Mars develop as a planet?
> -why is Mars different from Venus and Earth?
> -is there evidence of water on Mars, now or in the past?
> -what is the physical and chemical composition of Mars'

surface?
> -what are the characteristics of Mars' atmosphere?

It was important to stay longer with more sophisticated equipment and to send the type of spacecraft specially designed to search for these kinds of fundamental answers. With only a rudimentary knowledge of Mars as background, the Viking mission was designed to have two parts; a **Lander segment** to set down on the Martian surface and examine it, and an **orbiting craft** to map as much of the planet as possible. The requirements were tough because it meant designing two spacecraft to accomplish two much different tasks, and having them ride to Mars coupled together. And to meet budget constraints, both segments would have to travel on the same rocket and reach Mars at the same time.

To make matters a little more complicated, the engineering and data teams decided to build *two missions* – Viking 1 and Viking 2 – and launch them close to each other for an even more comprehensive coverage of the planet. The end product would be two orbiters and two Landers. Viking 1 Orbiter and Lander arrived in June 1976 at *Chryse Planitia* and Viking orbiter 2 and Lander 2 at *Planitia Utopia* in August 1976. Both Viking sets functioned normally and were able to carry out their missions without difficulty.

The end result for both missions was the discovery of a still unknown Mars, with some features for which scientists had no questions as yet. The missions were, as Dr. Young had predicted, a veritable feast of new information but once again the results raised a huge number of new questions about Mars.

As the craft reached Mars, the Orbiter powered up the Lander in preparation for separating. An aeroshell enclosed the Lander and bore the heat of entering Mars orbit. The aeroshell was let loose and at a specific altitude a parachute popped out to gently lower the craft almost to the surface. Retro engines then slowly settled the Lander onto the soil. The Viking Landers were miniature laboratories designed to dig into the Martian soil, dump the samples into a material analyzer, and extract information about the materials' composition. A meteorological boom sampled wind speed and direction, and determined the basic composition of the near-surface atmosphere. Cameras on the Lander gathered images of immediate surroundings as well as the functions of other moveable onboard scientific equipment. The data streamed in to the Landers' sensory equipment in a flood of new information.

Both Orbiters at the same time were continually capturing imagery of Mars' tortured surface with thousands of pictures providing glimpses of the planet's violent history and present condition. Complete analysis of the images took years, but the immediate first outcome told scientists that Mars as a planet was going to be difficult to decipher and that the planet was not going to give up its secrets without a fight. This was our first inspection of another world in space and the overall picture of Mars bombarded the science community with data that sometimes the scientists at first didn't know how to analyze. The science and technology of astronomy were catapulted into completely higher levels of professional practice.

This view of a model of the Viking spacecraft shows the white aeroshell underneath before it was released to expose the lander. Solar panels, TV cameras, antennae, and fuel bottles are exposed; in flight they are all covered.

Eight investigations were conducted from each Lander using specialized sensors and a surface sampler attached to a movable boom to dig up soil samples for incubation and analysis inside the biology instruments' three metabolism and growth experiment chambers. The Lander's cameras took pictures of Mars from the surface, and the folding meteorology boom took regular measurements of the near atmosphere, wind speed and direction. A seismometer measured seismic activity at each landing site. To gather up the data sent to Earth, the Viking mission was tracked by the NASA Goldstone Deep Space Network antenna. The Orbiter scan platform had a pair of high resolution, slow scan television cameras with filters controlled by ground command that scanned strips of the planet's surface. The Orbiter also looks for dust storms that might threaten the landing area around each Lander.

If there is life on Mars, the scientists figured, it most likely would be micro-organic. The three biology instruments were designed to detect microorganisms by analyzing metabolism and growth – changes by photosynthesis, or by the consumption and release of gases during metabolic change. Three special compartments on the Lander each processed a separate soil sample and tested for different properties; the tests took about 12 days to offer up the results.

Power for the Lander is mainly provided by two SNAP 19 style 35-watt radioisotope thermoelectric generators (RTGs) on top the Lander – in short, two nuclear reactors to provide electricity and heat to spacecraft instruments in the harsh Martian environment. In addition, each Lander has four nickel-cadmium rechargeable batteries that are charged by the RTGs. All this highly advanced, sophisticated equipment had to be built, tested and re-tested before installation on the Landers and Orbiters.

The Landers came down on large parachutes which are jettisoned at about 4,000 feet above the surface. At that time, the Lander is falling at about 138 miles per hour in the very thin Martian atmosphere. Soon the retro-rockets fired to slow the descent even more, and at last the Landers touched down at a gentle 5 miles per hour. The two Landers put down about 4600 miles apart on very different landscapes.

The huge amount of information learned by scientists at these two sites might very well be described as "the first day of class" at the Martian university. Once again long held theories were simply obliterated at a single stroke and replaced with new data for which there was no precedent. All we had known and were continuing to learn about Mars was of little value because each new exploration revealed a new dawn of a world that no one knew existed.

It is not the purpose of this book to explain in detail the many major features of Mars. Literally volumes have been written about the surface of Mars, but a few fascinating features are interesting to think about.

Mariner Valley with its dozens and dozens of side-valleys, that long stretching torn gap in the Martian surface that runs for thousands of miles, was a feature that took decades to decipher. The enormous expanse of America's Grand Canyon could be fit into just one of these little side valleys – and become almost totally lost. How did that rip in the Martian surface occur? What is still happening in Mariner Valley?

A scientist could devote a lifetime of study to this one feature and still not be certain the Valley was understood.

Valley walls continually collapse in a process known as *mass wasting*. A geologist once described the valley as a living entity that constantly renews itself as it changes shape. Winds that roar through parts of the valley can change the shapes of the walls and the valley floors by scattering debris in different directions or by sweeping the walls and floors clean of rubble. Geologic processes are ongoing, bringing large-scale changes without adequate explanations. Was the valley caused by earthquake – by one earthquake or several or many? What kinds of processes are going on deep inside the planet to cause such earthquakes? Is the valley new or old – and how old?

The southern half of Mars (below the equator) is pockmarked by thousands of large and small impact craters. Why is the southern half that has far fewer craters so different from the northern half? In a sense, the amount and location of craters on Mars tells something about the overall evolution of the planet just as the number and scattering of craters on the Moon tell us about that body's evolution.

A once little-known geologist named Dr. Eugene Shoemaker spent his life studying the process of *impact cratering* in the Solar System, including our own planet. Driven by questions he was unable to answer, he eventually formed an accepted theory demonstrating that at one time our Solar System planets all went through a long period of bombardment by debris that produced craters. His work finally was crowned with success by visible evidence of the famous Shoemaker-Levy 9 asteroid bombardment of Jupiter. The asteroid broke into pieces and collided with Jupiter before our very eyes as astronomers worldwide recorded the collisions on film. Now it is agreed among scientists that impact cratering is a major formative feature of planets in our Solar System. Earth has more than 138 known impact craters based on the research done by geologist Shoemaker.

The giant volcanic dome on Mars' northern hemisphere was first called Nix Olympia; the name was changed to Olympus Mons (Olympus mountain) when the spacecraft to Mars began to send back pictures of it. By close study, we realize the dome was created by repeated eruptions. On Earth, volcanic domes are created by the upheaval of magma from inside Earth's core. But because of the movement of Earth's tectonic plates, the domes are shoved out of the way after an eruption has occurred. On Mars, which is not so governed by tectonic plate movement, each volcanic eruption piles more and more debris on top the same dome until the dome rises to an enormous height.

Olympus Mons, located in the Tharsis region, has a 600 kilometers (372 miles) diameter and is 27 kilometers (16+ miles) high – higher than Mt. Everest. It is accompanied by three other massive volcanoes, all of them larger than most other volcanoes on Mars. The Olympus volcano has gone through several identifiable disfigurements as its walls collapsed or has been eroded. Also, Olympus shows evidence of impact cratering. It has taken a while to decide how these huge structures were created, changed and disfigured.

The Soviet record to reach Mars is not a good one. Of 16 attempts to explore the planet with spacecraft and robotics between 1960 and 1996, only 6 were "partially successful." This vacancy resulted in the United States being the only successful nation to study Mars close up. Since NASA's charter requires it to share space exploration findings with any country asking for it, the space agency supplied the Soviet Union with enormous amounts of data, some of it in real time.

America's successes have been because of exemplary management, but more so because of the amazing work of teams at the NASA CalTech Jet Propulsion Laboratory in Pasadena, California. Charged with carrying out America's planetary exploration programs, scientists, technicians and expert managers at JPL have always made the difference. Near the start of the Millennium, these experts inaugurated a robotics exploration revolution – and turned their attention fully on Mars. After studying for years, on August 6 1996 scientists studying a Martian meteorite found in Antarctica announced evidence of ancient life imbedded in the rock which measured 3.5 billion years old.

Pieces of surface rocks from Mars, like tektites from the Moon, are thrown out during the blast from an impact by a passing piece of cosmic debris or an errant meteorite or asteroid. This is a common occurrence in our Solar System. Once ejected by a collision event, the piece of Mars or the Moon finds its way to Earth where it is found and picked up by explorers and researchers. Microscopic images of organic molecules and fossilized remains in the Martian meteorite left little doubt in the minds of researchers

Officially designated by its name tag as ALH 84001 (for the Allen Hills of Antarctica), the announcement created a huge uproar in the scientific community that lasted for more than a year. Entire symposiums were devoted to discussions about the rock. As one might expect, the question of truth quickly divided the scientists into two opposing camps of thought. Prevailing wisdom today still asserts that the rock bears signs of past life on Mars as well as having been immersed in water.

Nearly a year later – on July 4 1997 – NASA introduced a new millennium policy and program of exploratory robotics it described as "faster, better, cheaper" methods of planetary study for Mars. The new program would focus on smaller hardware with single missions instead of the large, bulky and expensive spacecraft like Viking. On this date, a project called *Pathfinder* literally bounced across the red planet pursuing an impressive challenge for the little separate robotic land rover it carried. Using a "straight in approach" instead of orbiting the planet first, Pathfinder landed in Ares Valley adjacent to the remains of a large dry region the Mars scientists suspected of having been a lake or at least a channel for flowing water. Actually the Pathfinder mission was designed to be an *engineering demonstration* of key technologies and concepts for the future.

Pathfinder began its descent by parachute and braking rockets. A group of protective airbags completely surrounded the Lander and automatically inflated about five seconds before landing. The spacecraft hit the Martian surface at 40 miles per hour, bounced 50 feet into the air – and continued to bounce its way across the landscape 16 times before it stopped (about 500 miles from the Viking 1 landing site). The next day the rover rolled off its ramp and began its journey of discovery.

The robotic rover was designed to act independently and *to think for itself* in order to wind its way among the rocks and crevices.

A round-trip message from Earth to the rover and back again takes 24 minutes. If you send a message to the rover to "stop", it takes 12 minutes to reach the rover and another 12 minutes for the rover to tell Earthlings, "OK, I heard you." The six-wheeled rover has cantilevered arms so that the rover can climb over small rocks. The wheels have spikes on them to dig into rocks and the soil.

Rover returned 385 pictures of Mars landscape and determined elementary chemistry of surface materials. Rover simply butts up against a rock, pushes its APX spectrometer against the rock, and takes a radioactivity reading. Rover also has a laser capability that allows the little vehicle to search ahead and to each side for obstacles, and then make a choice to go around them or over them. In simple language: a hazard avoidance system.

The Pathfinder and its little rover vehicle were an unqualified success. They paved the way for future similar missions and gathered considerable data during their lifetimes. The official NASA Pathfinder internet website took an estimated 40 million "hits" on July 4, 1997 as Pathfinder landed on Mars, a sign that Mars is still an ever-present interest in the American culture especially. NASA plans to put astronauts on Mars by 2018 if all goes well. Mars will seldom be a final destination but rather a stepping stone out into the Solar System. And afterward? Who knows where next?

Mars at one time had oceans, lakes and flowing rivers. The landscapes were washed with rain and in wintertime there was an abundance of snowfall. Mars had to have been teeming with life of some kind. Of major concern among scientists is – what happened in Mars' history to turn the planet into an endless desert? Mercury, scorched by the Sun, is broiling with heat because it is so close to the Sun. Venus, covered with choking clouds of gas that created an unbearable greenhouse, keeps its surface temperature an inhospitable 900 degrees Fahrenheit. Of the four Inner planets, only Earth became a water planet. Why?

With the development of the NASA/JPL New Millennium Policy and its little band of Martian rovers, the study of Mars entered another exciting era that is producing spectacular results. In July 2007, the rover named "Opportunity" reached the impact crater Victoria after a 21 month journey across the landscape. Collaboration between NASA/JPL and Cornell University began a study of Victoria Crater because many of its well preserved features most likely hold clues about Martian evolution. Scientists began to give names to various parts of the crater, finally determining that it just might be possible for the rover Opportunity to descend into the crater and provide information about numerous features from a really close-up examination.

Maneuvering back to "Duck Bay", rover Opportunity prepared for its descent to the crater floor, but a dust storm began to develop and the scientists had to wait until it passed before initiating the tricky rover descent.

One special feature quickly captured the attention of scientists - a light-colored band in Victoria's wall at "Cape St. Vincent" that suggests a layering of material in the crater. Victoria is about a half-mile in diameter – not as wide as Meteor Crater near Flagstaff, Arizona – and about 200 feet deep. The rover was sent to the very edge of "Cape Verde" where it could peer down into the crater's floor and it was then that the layered light colored band was discovered.

5 THE BEST AND THE BRIGHTEST

"Really good teachers are not made – they are born. They can be trained and improved, but their value to themselves and to society depends on their innate talent and their belief in what they are called to do. This is a lesson administrators have never learned, and one most teachers forget all too soon. Exceptional teachers are desperately needed to teach in America's currently volatile environment. We cannot build a useful educational system on a diseased cultural foundation. Education as an American cultural imperative has to show up on someone's list of priorities."

> Tom Becker, International
> Space Technology Educator

Students at the British Space School ponder over a satellite problem in 1989. Requirements for acceptance to the summer school were very stringent to assure only the best students were accepted.

Teaching is fifty percent subject matter and fifty percent presentation, or "showmanship" if you prefer. Although the goal is to present the subject matter in such a way that your students can understand it on an intellectual level, of equal importance is a teacher's ability to get students to want to absorb it on an emotional level. The teacher's love of a subject – his/her own excitement and passion for the subject and desire to share it with others - are priceless qualities that make one teacher stand out above all the others. Most of my special teaching experiences were outside the normal boundaries of standard curriculum. In the practice of classroom technology teaching, opportunities led me to 45 of the 50 states, 12 countries of Western Europe, and presentations to the British Association for the Advancement of Science.

The opportunities to teach subjects they know well, to students who have a love of learning in the first place, made all the difference in the many special teaching appointments I was privileged to be offered. One other factor that *must be part* of the teaching equation is the courage and intelligence of administrators to turn deserving teachers loose and let them execute their craft as teachers and develop their own style of presentation. Really good teachers will excel in this kind of environment to an unimaginable high level; it's what makes a born teacher an outstanding teacher.

In general, the United States Constitution is mostly silent about responsibility for public education. It only says that powers not enumerated in the Constitution are left to be exercised by the individual states. In America, each state has a state department of education headed by a Commissioner of Education and directed by a state Board of Education. These bodies determine the kind and extent of public schools as well as the subject matter in the curriculum. That is one of education's major problems – we have fifty different kinds of education and fifty different requirements for high school graduation and fifty different sets of requirements for hiring teachers. Young people graduating high school in Vermont may not receive the same quality of education as students in Iowa or Texas; in fact, students in these states may not even have the same or similar kind of subject matter to study. There are no "standards" in professional education.

Unfortunately – VERY unfortunately – a teacher certified in one state cannot necessarily teach in any other state. The State of Pennsylvania, for example, will not allow teachers from out of the state to teach until they have fulfilled a long list of new qualifications. This is a politically-motivated requirement – and the cost to the teacher is absolutely outrageous, often amounting to hundreds of dollars and requiring many months or years of new enrollments in university classes they don't need in order to accomplish a new state certification.

The United States Department of Education, under the Executive Branch of our government, has absolutely no jurisdiction over the American educational system. The department is a research office that conducts studies of education across the country but without the power to bring about change. It can *recommend* levels of curriculum, *suggest* projects and programs, and *encourage* schools-teachers-administrators in various ways; *but it cannot mandate change.*

The American educational system is a self-defeating concept despite SAT scores to the contrary. Study after study of America's public school systems have accomplished little or nothing to advance the practice and organization of professional public education. *The National Defense Education Act of 1958*, a reaction to the Soviet launch of Sputnik the year before, and the publication in 1983 of *A Nation At Risk* by the National Commission On Excellence In Education, have brought no appreciable change in the public education institution in America in the past fifty years. Like a number of other professions, the institution of education is archaic and disorganized, and the future citizens of our nation are suffering as a result.

Teaching computer familiarity in the classroom is not "technology education." It is no different from typing classes or biology laboratories. A teacher in a southern Midwestern state called me one morning in the late 1980s. She literally was in tears and broke down several times during our conversation, out of enforced frustration. The superintendent of her school district mandated that every teacher must have a computer on the desk; the computers were furnished by district funds. "But we don't know how to use them," she explained, "and we don't have the time to learn how to use them in the middle of the school year." How does a man like that get to be a superintendent? How do teachers contend with students who are highly skilled in computer literacy, when the teachers are still in the dark ages of computer communication and technology?

A highly unpopular superintendent of a school district in Wyoming in the 1980s engaged me to talk to the District's teachers as an opening event to the coming school year. As an aside, the teachers told me they had car-bombed his car one morning at the close of the previous school year. Nevertheless I agreed and spoke to an assemblage of two thousand teachers about the emerging global culture of technology. Afterward, a teacher committee recommended that I be hired as a consultant for the coming year. The superintendent refused and confessed he could better use the funds to purchase new furniture and curtains for the teacher lounges.

I was teaching high school world geography in an affluent St. Louis suburban school district in the 1980s. To make geography more meaningful, I used satellite imagery to supplement textbook maps. At my own expense, I was supplying students with printed images in color. One afternoon at the end of the day, an assistant principal stopped by my classroom. Putting a "fatherly" hand on my shoulder, he said bluntly, "You will never get anywhere in this district if you keep teaching this space stuff." Taken aback, I asked why. "What good is teaching space to these kids instead of teaching real geography?"

In the 1980s, I taught high school in two different integrated school districts amid some of the most threatening environments of my career. In one district, while I was teaching cultural technology classes in summer school, I heard a consistent bump-bump in one of the student desks. Experience led me to believe there was a gun in the room and I called for the Principal and a Security Guard who lined the students around the room to frisk each. Unknown to the authorities, the students were passing the gun back and forth behind their backs. The weapon was finally found inside a covered radiator. In another instance I called for the Security Guard when a student ran outside my first floor classroom firing a pistol. Two of my students that year arrived for final exams on crutches and bandages from being shot in the legs the night before.

During my career, two of the principals I worked under were arrested for embezzling school district funds. One of them was quickly hired as a lead principal in a highly affluent Colorado school district; the other was placed on court-mandated probation. And finally, the State Commissioner of Education was arrested and convicted of shoplifting from a local liquor store. On the last day of a school year in one district, I was summoned to the office of a Principal I considered not very bright and told bluntly, "This is the bottom line for you. You will stop augmenting curriculum with irrelevant subject matter or you will no longer teach here." Naturally I resigned and went to England to teach technology to pre-university students.

The students I was teaching, from 4th grade through Graduate Teacher Symposiums, belonged to a category that is professionally known as *Gifted*. The students exhibited, and were tested for, higher level characteristics of exceptional memory, consistent curiosity, personal commitment, advanced language skills, personal organization, high general intelligence, excellent creative skills, strong aptitude for problem solving, acceptable social skills, exceptional motor skills, and advanced desire to succeed. Classes filled with students like these are lively, animated and frequently humorous. Many times I was backed up to the chalkboard with questions I had never even thought about. I often learned as much from my students as they learned from me because we ended up teaching each other.

An 11ᵗʰ grade student in a 1998 Missouri Scholars Academy class "Landscapes Of Mars" studies images of the Martian surface captured by the Viking Landers.

The schools I taught in were extra-curricular, meaning outside the normal standard public schools. They were taught in the summer or on weekends or after normal school hours. Classes usually consisted of problem solving exercises that required students to think at higher levels and to learn from materials for which they had to find their own solutions. The main focus was on critical thinking and personal creativity.

One of the favored student exercises was to be presented with a color image of the volcano Olympus Mons on Mars which has an exceptionally clear view of the volcano's caldera (sunken circle at the top) showing evidence of six separate eruptions. The students had to determine the order in which the eruptions occurred, make a tracing of the caldera and label the parts in the proper order of eruption. Then they had to write a summary stating 1) how the image was obtained, 2) the height, diameter and other measurements of the volcano, and 3) any unusual characteristics of the volcano.

Another favored exercise was to examine a classic Landsat satellite view of The Netherlands, identify and label all its geographic parts, and write an explanation of how the Zuider Zee was changed by construction into the Ijsselmeer. Or – using an aerial or satellite image of Meteor Crater in Arizona and an internet-connected computer, write an explanation of how the crater is believed to have been created with a tracing of the actual crater. Or – using one's own diagrams and sketches, explain how the Thames River Flood Barrier System operates to prevent the flooding of London in an anticipated severe storm. Or – explain how the Chesapeake Bay Bridge Tunnel was constructed using carefully drawn diagrams and sketches of construction methods and a sufficient amount of original text.

The list goes on and on, usually ending with a final whole-class project showing and explaining how a simple research station on Mars or the Moon could be built. The class is separated into research groups, each of which researches a section of the completed project. The final report is contained in a booklet, composed on the computer by the students, with appropriate sketches and/or diagrams of the students' own creation. This project determines if the students can be successfully graduated from the class. The higher the students' age level, the more professional their final product needs to be and the more difficult the project they are working on.

Each of these exercises taps into many of the gifted student's higher order of abilities and always *requires them to think on a higher level*. Thinking is a skill that must be continually developed in gifted young people; higher order or imaginative, visionary thinking usually has to be trained. Most young people think sometimes about all sorts of daily and "futuristic" topics, but usually in a disorganized manner. Higher order thinking already is present in the gifted child – it just isn't focused or developed. Exercises such as the ones described here are definite goal-oriented thinking exercises.

Space technology, as daily instruction, cuts across all academic disciplines: astronomy, physics, chemistry, English composition, English literature, geography, geology, aerodynamics, psychology/sociology, music – even cooking and food preparation. It therefore usually is

-interdisciplinary
-extra-curricular
-multi-cultural
-and reality-driven

In the gifted space classes I taught, we often studied all these subjects simultaneously for a single project. Think about it - if the focus of a project is building a research station on Mars, we will have to use all these subjects.

There is one more dimension to teaching gifted students; they don't want to think of themselves as "different" or unusual in any way. Students in both Scholars Academy and British Space School classes continually protested against the "gifted" labeling of their abilities. In most cases, they were required also to attend classes in social behavior because they usually acted or spoke differently than other young people. Their behavior often brought them into conflict with teachers and other students because the subjects they thought about were out of the realm of more usual behavior.

If you read Book 1 of the Technology Trilogy *(Eight Against The World),* you already know that those valiant warriors constantly were at odds with society, neighbors, government, and certainly the established Church. The exact same problem exists today for gifted students. Poor Johannes Kepler was always getting beat up by his classmates because of his perceived arrogant and know-it-all attitude. Isaac Newton was shunned by his classmates because he was just "different" from them and they didn't understand him.

Each time I began a class of gifted students, I tried to invite their parents to a one-hour meeting to explain how the class works. It is highly important to have parental support because a group of highly motivated, highly creative students can also have high expectations and want their parents' help because the students are going to be asked to do things they most likely have never done before. Parents need to know what these things might be and why they are being asked to help with them. In many instances, when the students went back to their schools they did so with knowledge beyond that of most of their teachers and that kind of situation can cause all kinds of social and academic problems.

Walking through the hall of the high school where I was teaching on a daily basis, I came upon a bulletin board poster early in 1985 with the following announcement.

WANTED: OUTSTANDING TEACHERS TO APPLY FOR FACULTY POSITIONS IN THE 1985 MISSOURI SCHOLARS ACADEMY
Sponsored by the Missouri State Legislature and the Dept of Elementary and Secondary Education and the University of Missouri at Columbia

I didn't consider myself to be an "outstanding" teacher, but I thought it would be an opportunity to teach technology. Responding to my application, the review board called me for an interview. In the course of conversation, one of the Academy Directors said, "Your letter was so arrogant and demanding I just had to meet you." After that I was accepted.

The students were from all over the state – more than three hundred of them to attend a three-week on-site enrichment program. Choices for student enrollment ranged from communication and mathematics to science, social studies, philosophy and Interpersonal Skills. Students lived in dormitories and ate in the University cafeteria. Faculty housing also was provided by the University. Funding was made available by the Missouri State Legislature, the State Board of Education, and the University of Missouri.

Students were required to complete a rigorous application process that gave evidence of a strong academic background, community involvement with written recommendations and evidence of some successful extra-curricular activity. Needless to say, the students enrolled in my classes were outstanding scholars from every imaginable large and small town. They also were everything meant by the term "gifted."

Teachers were well paid for their services. I taught a major course *Science and Technology in Space* that included reviews of the world's major space-faring nations, the major space missions with emphasis on European and Soviet achievements, and a minor course in *remote sensing technologies by satellite.* Classes were conducted in the morning, special lectures and presentations were held in the afternoons and evenings. Students also participated in other activities such as a student newspaper, personal hobbies, and short special academics. Class sizes of 25 or less students were the norm.

Student experiences with space technology were mostly limited to computer games, sci-fi movies, and news broadcasts and telecasts. They had little or no background in space affairs or general geography, but that isn't surprising. The State Board of Education for Missouri in 1986 recommended removing geography from the required curriculum – a totally unthinkable decision considering America is involved in a global marketplace with space-based technology and spread out over geographically remote locations and countries.

Addressing a large gathering of a local Mensa Society chapter in the mid-1980s was a major wake-up call in the midst of my career. The huge gathering represented the top one percent of the population in measured intelligence, yet the subjects I was talking to them about were, for most of them, a completely new introduction. My talk was about satellites and global geography accompanied by nearly one hundred color slides.

After the talk a young woman came up to me with tears rolling down her cheeks. "I just loved your talk, and I was fascinated by your pictures, but there was no geography course in my school district – and I feel cheated!" How long will it take for educators to wake up to the reality of the Global Village?

The Academy called me back for the 1986 session to teach more space science. In 1998 I again applied for and won a faculty position to teach about the planet Mars. I also had the chance to give a special student seminar about commercial space art in which the students used their own art materials to create some excellent visions of the space frontier. Always an exhilarating experience, I once again enjoyed conversations with students and other faculty members and the chance to speak about the human reach into space.

One afternoon in early 1989, the phone rang and a very British voice asked to speak to me about doing some teaching at the British Space School just being formed at Sevenoaks, Kent in southern England. It was the answer to a silent prayer that eventually led to so many other opportunities it is difficult to remember all of them. All that I had ever read about, dreamed about, studied about, yearned to see, of European history and culture – suddenly was all there in front of me; architecture, paintings, colleges, cathedrals, public school systems, transportation technology.

Once in Europe, I ran my fingers over the Rosetta Stone, stood in awe before the Mona Lisa and the Venus de Milo, walked inside Westminster Abbey before the tomb of Isaac Newton, gazed in wonder at incredible Wells Cathedral, feasted my eyes on Shakespeare's home, passed through Swiss mountains and the walled city of Lucerne, gazed in amazement at the mysterious moors of Cornwall and Devon, photographed the Night Watch in the national museum in The Netherlands, and visited the great brooding stones of Stonehenge. Here were the roots of the English-speaking world and my adventure became a pilgrimage into the past.

Students at the British Space School were called Sixth Form students – between high school and university – searching for careers, for that one brilliant spark that might determine the rest of their lives. They were preparing for their exams and their original research projects that might guarantee them admittance to Oxford or Cambridge or some other university they longed to go to. Without a doubt they were gifted and I relished their amazing questions and their point-blank attentiveness.

After five years of teaching at the Space School, first at Kent and later at Brunel University in Uxbridge in West London, we learned that our students went on to amazing careers after university, many of them capturing national awards in geography, mathematics, history and science. I came to believe that, while American students are fairly well schooled, European young people are very well educated. The Space School was funded by aerospace companies eager to hire bright young students after university, and partly by the Brunel Bioengineering Institute at Brunel University. Parents paid tuition fees or else the students worked at local community projects to earn the tuition themselves.

Mind contests, building and launching model rockets, assembling radio transmitters, studying satellite images, working out problems in astronomy and physics, and learning – learning – learning. Corporate industry donated computers and funds for teaching materials, sent special exhibits and mock-ups of actual space hardware such as satellites and aerospace planes. On field trips, students participated in special investigative tours of universities where astronomy and biology and mathematics were taught. Although the live-in sessions were only ten days long, young people from various parts of the British Isles got to know each other quickly; young people are all alike the world over.

The Space School eventually ran out of funds and had to close its doors. The letters I received from former students, however, told me we had struck a chord that changed their lives. On the last day of the last class of 155 students, I stood at the podium in the lecture hall and looked across the mixture of incredible young people. Silence came over the group as they waited for some final words of great wisdom that I couldn't seem to pull together.

In a few moments they would be out the door - gone forever. What could I tell them to send them out into the sunshine filled with new hope for the kinds of futures that were theirs for the taking? Should I tell them they would have to fight the rest of their lives for what they wanted? Remind them that learning is a lifelong task? No – I needed for them to bask in the great promise of personal exploration because they are witnesses to the birth of humankind's first feeble steps into an endless, unforgiving universe.

"In order to achieve all that is possible," I began, using a quote from Gale Baker Stanton, *"we must attempt the impossible. To be all that*

we can be, we must dream of being more." Suddenly what I needed to say fell into place. *"You must be willing, at a moment's notice, to give up who you are in order to become who you can be. The space frontier does not call out to the faint-hearted, does not sing its siren song to those who waste themselves and choose not to achieve something of value. Don't limit yourself until you find out what your personal limits are.*

It is the fear of change, not the activity of change, that keeps us from achieving the progress we so desperately seek. We must never close our minds to the great ideas of our time, for that would be to turn our backs on the wonderful magic of the human imagination. There are new worlds out there, not to conquer but to discover; not just to explore but to celebrate. Go – take up the outward journey – and learn.

By 1994 I was beginning to feel the first indications of later adulthood - beginning to enter the twilight of my life. The long flights across the Atlantic and back and forth across America, the lack of sleep in order to meet deadlines and create teaching materials, the endless days of always being on parade in front of audiences of students and adults, the timeless march of "stand and deliver" to tell, to inspire or to coax, were beginning to take its toll.

Heaven had watched over me – had guarded my health and led me to these far off places and filled my life with the beauty of music and art and learning and watching young people "become." Now I just had time to catch my breath before beginning again. Somehow, sometime, the phone would ring once more and an unfamiliar voice would say, "I'm trying to locate Tom Becker the space educator. Are you that person?" But until then, it was time to sit back and take a breather to get ready for that next phone call.

At the start of the 3rd millennium, we are all "on call." Someday we will be asked to give the best that is in us, to reach beyond, to search inside ourselves and come up with "it." Somewhere along the way an older and wiser man I admired once told me, *"At the end of the day, when you are exhausted and just plain weary, you must be able to look in the mirror and say that you made a good accounting of yourself - that you gave the best you had in you."*

Vince Lombardy, the rugged coach of the legendary Green Bay Packers football team, perhaps said it best another way. "Winning is not a sometime thing; it's an all time thing. You don't win once in a while; you don't do things right once in a while, you do them right all the time. Winning is a habit."

By the 1980s, all human knowledge DOUBLED according to a prescribed equation that is breathtaking in its scope. The fast pace of the knowledge explosion has had enormous impacts on our daily and professional lives. In the year 1975, all knowledge DOUBLED every ten years; in 1985 it DOUBLED every 20 months; in the year 2000 it DOUBLED every 10-14 weeks.

By 2010, all knowledge will DOUBLE nearly every week. Difficult to grasp on an emotional level, we wonder what amazing things will be discovered before this century is ended. In the year 1000, humans in Europe in tiny villages were cooking over outdoor fires, living in huts and lean-to shelters, and killing each other in tribal wars. In a thousand years of "progressive" humanity leading to the year 2000, air conditioning, refrigerators, automobiles, hair dryers, cell phones, watching movies in our living rooms, restaurants and convenience stores, and global passenger airliners were taken for granted. What will we be taking for granted at the start of the next millennium in the year 3000?

6 INCIDENT AT CHERNOBYL

"An accident has taken place at the Chernobyl power station and one of the reactors was damaged. Measures are being taken to eliminate the consequences of the accident. Those affected by it are being given assistance. A government commission has been set up."

Terse Statement by Radio Moscow, Official Soviet Government, April 28 1986

Chernobyl nuclear reactor 4 after the explosion showing extensive damage to the main reactor hall (image center) and turbine building (lower left). Photo courtesy Wikipedia Encyclopedia.

The vast expanse of the Ukraine region in the former USSR is the Soviet Union's "bread-basket" where enormous amounts of wheat and grains are grown to feed Russia's millions of people and millions more in the outside world. Huge pieces of farm machinery from tractors to columbines and thrashers ply the seemingly endless fields often ten to twelve hours a day. Farmsteads dot the countryside for hundreds of miles in every direction that look not much different from farmland in any large country. For the Soviet people, the region offers a good life for raising a family and a broad selection of grazing stock and barnyard animals.

On April 24 1986 a verbal order was given to the technicians and managers at the Soviet nuclear power plant at Chernobyl, a small town in the Ukraine region just north of Kiev, to conduct a scheduled equipment test. According to safety requirements, a test of the nuclear plant was required as soon as possible to re-certify that the plant was a safely operating facility. The test was not completed because of an explosion. In fact, shortly after the test began, a catastrophic incident occurred that would destroy this idyllic region and suddenly cause its usefulness to disappear overnight – perhaps for the next hundred years.

At 1:23 in the morning on April 26 1986, the required test on number 4 reactor was underway. Because of human error, there was a buildup of steam that caused the reactor core to get out of control and an explosion took place that rendered the reactor useless. From this point onward, the situation at Chernobyl and the neighboring town of Pripyat continually worsened in stages. The roof was blown off reactor number 4 and instead of a meltdown the core exploded showering the landscape with radioactive debris and setting fire to the adjacent building housing reactor number 3. The prevailing wind was light and the smoke and radiation were carried northwest from the towns of Chernobyl and neighboring Pripyat where the nuclear plant workers' families lived.

The plant at Chernobyl underwent several standard tests of failsafe systems. Some devices didn't work properly. Suddenly there was a loss of cooling system to cool the uranium fuel rods. The disaster was brought on by a fuel rod "meltdown" leaving the technicians unable to control the nuclear furnace. The result was a core explosion; the amount of radiation leaked was a hundred times greater than the amount released at either Hiroshima or Nagasaki.

The force of the blast from reactor 4 carried chemical pollutants straight upward and at a very fast rate - cesium 137, strontium 90 and iodine 131. Chaos took over immediately as plant evacuation horns and emergency sirens sounded, and the undulating wail of fire protection sirens. Fire rescue teams rushed to the scene at reactor 4 but were warned away because of radiation. However, firemen still tried to put out the fire in the reactor core but without success. Of the first 31 casualties at the plant site, 27 were firemen. These 27 firemen and the remaining four all died within months of the disaster.

Work crews were called to the scene, mostly miners. If the core burned through the flooring, or the weight of the debris that fell on it was too heavy, it would crash down onto the floor below causing more explosions and spreading still more deadly chemicals. To prevent this, some 400 workers in three shifts worked around the clock to build a platform beneath the reactor core. Families of plant workers were not notified for twenty-four hours. The morning following the accident, younger children were outside playing and older children went to school as usual so that everything gave the appearance of normalcy. Soviet officials explain that there was no early warning system set up. Soviets are fearful of public panic and they will not tell the public about an emergency until they understand the situation no matter how long it might take before all the facts are in and the puzzling "situation" becomes understood. Even then, so the record shows, there might be continued delay in public notification. In minor disasters, the delay might not cause immediate threat to the public; in the Chernobyl incident the delay proved to be highly life-threatening and devastating.

Map of Russia

After thirty-six hours, when it became clear radiation levels were rising and not dropping, government officials began a full evacuation of the population of Chernobyl and Pripyat. The task was completed in about three hours. By April 28, technicians at Sweden's' Forsmark Nuclear Power Plant 60 miles north of Stockholm monitored high radiation levels and began checking with nuclear plants all over Europe. Radiation was present over all of southern Scandinavia and the official government representatives in East and West Europe countries suspected the Soviet Union.

The traditional Soviet bureaucratic "wall of silence" denied anything was wrong although American and French reconnaissance satellites operating at about 150 miles in space already were starting to get high levels of radiation readings. Working closely together, American and European allies finally located the source of radiation at Chernobyl and demanded a statement from the Kremlin. At long last, Radio Moscow issued a short, terse statement that owned up to the overall situation but provided little or no details and no hint of the seriousness of the incident. The statement was perilously close to a denial of the real severity of the event and far from "forthcoming" as the rest of the world had hoped it would be.

"An accident has taken place at the Chernobyl power station and one of the reactors was damaged. Measures are being taken to eliminate the consequences of the accident. Those affected by it are being given assistance. A government commission has been set up."

Three foreign intelligence satellites were pressed into service to locate the source of radiation, all of them looking at the Soviet Union since the overall situation had all the earmarks of a Russian cover-up. The general population had little or no knowledge of the capabilities of these satellites since most of their work was kept quiet. My own studies of satellites led me immediately to track them and the specific imagery they developed.

The most obvious first satellite choice was the American Landsat satellite, an Earth Observing bird with broad abilities designed to look at any kind of geographic feature except water. Landsat is not designed to discern water features. The second choice was the French SPOT satellite, much like Landsat except its optics are somewhat better. Landsat was and still is considered the "workhorse" of the American satellite fleet.

The third satellite, however, is cloaked in deep secrecy and referred to simply as Keyhole – a dedicated satellite used exclusively for clandestine observation (spying). Keyhole had gone through several stages of development and could literally read a license plate from 150 miles in space. Operating between 150 to 250 miles in orbit, the three satellites, if trained on the same target, could identify a bird sitting in a tree and determine what color it is.

I was attending an informal reception one evening in 1984 and was drawn into a conversation with an army Colonel talking about American capabilities. He described our satellite work missions this way. (My own opinion at the time was that the Colonel had shot his mouth off and divulged more than he should have, but he was in a bragging mood and wouldn't be deterred.)

"If you're having a backyard dinner," he began, "and the tables are laid out for your guests, we can tell how many people are invited to the affair. We can tell if you're serving cantaloupe or watermelon.

If your guests have arrived we can separate civilians from uniformed military personnel and for the most part tell what their military ranks are. By calculating the height of the military personnel, we usually can tell if they're Asian or Western. We can find out exactly who everybody is by reading the license plates on their vehicles, and we can tell by reading the heat reflected from their automobile hoods pretty much the exact order in which they arrived at your house. If they've flipped a half dollar coin to decide who is going to say the prayer before dinner, we can tell if the coin came down heads or tails."

We were civilian guests in 1977 at a briefing by an officer of the Strategic Air Command who winced at some of our point blank questions about the use of satellites. For quite a while he kept silent, admitting only that we had satellites. Since I had been studying satellites for some years I was able to discern what he *wasn't* willing to talk about. We quickly proved to him we were a satellite-wise group and to stop stalling. Suddenly he astonished us by blurting out, "We have reconnaissance aircraft skirting the Soviet Union about 60 miles out from their borders – and we can count the slats in the venetian blinds in the windows." That one piece of information told me that our optics had just been upgraded in the Keyhole satellite series.

The search for and assessment of the radiation source at Chernobyl didn't take very long – maybe about 12 hours altogether to get the satellites into position. Looking for heat surges and discolorations at known nuclear reactor sites, the culprit was quickly identified and permitted an assessment of how serious the situation was. American doctors began volunteering to go to the Soviet Union to help with victims; they were immediately denied entrance into the country and told that the Soviets didn't need help because their own doctors were very capable.

The American media described the Soviet dilemma in characteristic style. By then the prevailing winds had shifted and were blowing the chemical-laden clouds towards the east. Governments of countries in that direction took immediate precautions:

Poland banned the sale of milk from cows that had eaten fresh grass; children from birth to 16 years received iodine solution injections;

Rumania declared a state of alert and warned people to stay indoors and not drink water;

Austria told pregnant women and children under age 6 to stay indoors; outdoor vendors of vegetable stands were told to cover their produce; etc.

At Chernobyl, Pripyat and Kiev, radiation contamination entered the food chain and the water supplies including lakes and rivers; fields of crops, pasturage for livestock, and water reservoirs. Pripyat, the small town created 3 miles from the plant for workers and families, is beside the Pripyat River. The Chernobyl plant draws water from the river to help cool the reactors. Since the force of the blast was straight up, authorities mistakenly believed the townspeople were in no danger at first and that is why no one told them what had happened.

At the time of the explosion, 3000 REMS per hour - enough to kill someone in less than 20 minutes – were released into the atmosphere. The contaminated clouds reached as far north as Finland and as far south as Greece, Turkey and southeast France. The first figures of casualties released by the Soviets showed a continuing upward climb. By May 2, two deaths were reported and 197 hospitalized; by May 14 there were nine deaths and 300 hospitalized. But these figures centered only on work crews and of course had nothing to do with the population at large.

Immediate effects of radiation poisoning are painful and slow:

-heavy depression of white cell count
-severe intestinal syndrome in 1-4 days
-considerable loss of hair
-nausea and vomiting

Reactor 4 was completely sealed over and around in concrete and a concrete platform was built under the burning reactor. All materials (trucks, cars, tools, etc) were buried in pits lined with concrete. Trees, crops, grass, dirt surfaces all were plowed under – a useless gesture since radiation still was free to come back to the surface.

Twenty countries were affected by the Chernobyl fallout. Laplanders were hardest hit, mainly because it rained the day after the Chernobyl incident causing thousands of reindeer to be killed. In all, 400 million people were affected by the Chernobyl disaster. All fifteen of the Soviet RBMK reactors were shut down and overhauled. The Soviets use a mixture of graphite and water to cool the fuel rods (Americans use water only). This construction is looked on as a major weakness in the design of Soviet nuclear operating plants.

Because of its commitment to a national policy of isolation from the rest of the world, the design of the series of nuclear reactors built in the Ukraine was originally flawed. Chernobyl was the first operational Ukrainian reactor, commissioned in 1977. Together with insufficiently trained personnel who made major operational mistakes in 1986, and the absence of a cultural emphasis on safety features, the Chernobyl incident was inevitable. The Ukraine operates 15 reactors today, all built on the same RBMK design principle. After the Chernobyl explosion, all these reactors were quickly upgraded and re-engineered so that a similar explosion couldn't occur at another location.

It was officially reported by the Soviets that nearly two thousand babies born after the Chernobyl incident by mothers exposed to high doses of radiation were delivered in healthy condition and with no side effects. But in such places as Hospital #6 where exposed workers were taken after the event, patients continued to die from radiation effects. Some Soviet officials describe the radiation sickness as a slow and long-term disaster. The 1,959 babies born from women at Pripyat at the time of the explosion have proved normal *according to official Soviet reports.*

Neutral observers, however, have a far different story to tell. *Chernobyl Children Life Line*, a British charity organization located in the Guernsey Islands off the coast of France, reported in 2007 that:

-2 million people live in radioactive contaminated territories
-only 5% of children who lived in the Chernobyl area at explosion time are healthy
-there is a 40% increase in malignant tumors in men, 29% in women
-there is an alarming increase in tumors of bone marrow, the brain, mammary glands, kidneys, lungs, bladder - and thyroid cancer

People living in contaminated areas drink and bathe in contaminated water, eat food grown in contaminated soil, and are subjected to still-radioactive environments. More than 300,000 people are dying a long, slow death that is not reported in the media. A casual browsing of *Chernobyl Deformities* on the Internet is an international wake-up call for those of us who aren't aware of the horrible consequences of the Chernobyl event.

One of the most glaring mistakes to come from the Chernobyl incident is the failure of Soviet authorities to take immediate action when they learned of the explosion until 48 hours after the accident occurred. Mistakes in judgments of the seriousness of the situation allowed millions of people to carry on their daily activities without any knowledge of the life threatening incident and no immediate warning to take shelter or evacuate.

In 1987, Andrei Sakharov, "father of the Soviet hydrogen bomb" and strong human rights advocate, was released from seven years of house arrest along with his wife Elena. On March 16 of that year at age 65, he startled the world by showing up at a Moscow forum on nuclear energy and delivered three speeches in which he severely criticized the Soviet government for its cover-up of the Chernobyl incident.

People around the world still continually ask, "What did we learn from the world's worst peacetime nuclear catastrophe?" In summation, we learned a great deal about human exposure to direct radiation poisoning and how to treat it; we learned how important the right kinds of safety measures are at nuclear plants; and we learned the importance of constant training and retraining of plant employees. Safety and safety training must be priorities, but in the Soviet Union of the 1980s the value of human life was negligible as a cultural imperative. Proper construction of nuclear operating plants in America is a first priority requiring constant inspection and approval by the NRC.

In America, the Nuclear Regulatory Commission works very hard at preventing the kind of incident that occurred at Chernobyl. At plants like the Limerick nuclear operating station in suburban Philadelphia and the nuclear plant near Fulton, Missouri can boast of more than twenty years of operation without a single safety violation.

Frequent inspections, continual training and updates about nuclear equipment and mechanics, and a strict compliance with NRC rules and regulations often result in more than 20 years of perfect records of safe daily plant operation.

Eventually the Soviet government relented and allowed doctors and other medical personnel to attend to patients and perform treatment programs. More modern equipment from United States medical organizations as well as some other equipment from several other countries replaced old and failing hospital diagnostic equipment in addition to treatment packages and medicines. Doctors who responded to Soviet acceptance of help described Soviet medical technology as "far outdated and inadequate." Personal working relationships between American and Soviet doctors and patients resulted in a number of lasting professional friendships.

Long term illnesses from radiation poison continued after many years even into the new millennium. Occasional scheduled checks for background radiation are conducted by Soviet authorities that indicate a continual but slow drop. In some areas people were allowed back into once-contaminated areas, but for the most part inhabitants only returned to reclaim personal belongings and did not stay to take up their old residencies.

In some nearby areas, clothes still flap in the breeze, doors are left wide open and children's toys remain strewn about yards. Even in areas considered now "safe", there is a landscape of empty dwellings and a surrealistic silence of emptiness that people from normal cities cannot abide. The streets are empty, buildings are overgrown with vegetation and cars and trucks still sit where they were abandoned. And it will continue to be this way for dozens of decades to come.

In America, the only agency that stands between the population and nuclear plants and a Chernobyl-type incident is the Nuclear Regulatory Commission. The public and cultural responsibilities of the Commission are truly awesome. Because of the agency's constant vigilance, there has not been one single Chernobyl-type accident in the United States (or anywhere else in the world).

The behavior of Soviet officials in positions of authority laid bare the dysfunctional nature of the Soviet system of dealing with problems. With the lives of literally hundreds of thousands of citizens at stake in a fast-moving catastrophic incident, officials still were unable to think in terms of human safety. At that moment, the single over-riding factor was concealment and protection of the Soviet system NO MATTER WHAT THE COST to the population.

It is alarming, by western standards, that government officials in charge of nuclear energy as a daily working commodity didn't understand the very subject of their work enough to realize how extremely serious the situation was. At the same time, officials *failed to demand* that plant employees were continually trained, re-trained and tested in such concepts as specific uses of equipment, theoretical nuclear physics, development of energy from nuclear sources, failsafe systems, emergency procedures, and the like.

The greatest single failure in the Chernobyl incident may be summed up in a single sentence: there was a failure at the highest level of government and common sense management systems which caused the highest catastrophic level of human suffering across a territory of 20 European nations. There were so many failures in the Chernobyl incident it probably is not possible to identify all of them!

7 ONCE IN A LIFETIME

"I came in with Halley's Comet in 1835. It's coming again next year; it will be the greatest disappointment of my life if I don't go out with Halley's Comet."

Mark Twain, 1909

Gas spews forth from the nucleus of Halley's Comet in this image returned by the European Space Agency's Giotto spacecraft. Photo courtesy ESA and Max Planck Institute, Germany.

By any standard, Edmund Halley certainly was not an ordinary man. Known mainly throughout history as an astronomer, he also did pioneering work in ocean navigation, tides, the trade winds, cartography, mortality tables, and two proposed designs for a diving bell. Although he never lived to see the full results of the work he did, predicting the period of the comet that bears his name along with his other many accomplishments in astronomy won fame and fortune for him during his lifetime.

Edmund (also spelled Edmond) was born on November 8, 1656 at Haggerston, Shoreditch, England close to London. His father was a successful merchant in the business of making soap. The young boy was educated at home and immediately took a liking to astronomy which eventually became his life's work. Born in the flowering of the Renaissance, he was encouraged in his studies by his parents and was privately tutored, learning at an early age all the rudimentary concepts of astronomy and astrology. At age 17, he entered Queens College Oxford already with an expert background in astronomy, bringing with him the best astronomical instruments of his day provided by his wealthy father.

It was natural for Edmund to become allied with the Royal Observatory at Greenwich in 1675 assisting John Flamsteed, the Astronomer Royal, with astronomical observations. The following year Edmund went to St. Helena, Britain's farthest territory in the southern hemisphere, where he catalogued 341 stars and 24 comets over an eighteen month period. He returned to Oxford to publish his findings in a catalogue and in 1678 was graduated without having to take the examinations on decree of King Charles II. He was elected a Fellow of the Royal Society, a group of the foremost scientists of his day, as one of its youngest members.

In 1682, Halley went to visit Isaac Newton because he (Halley) was having mathematical problems with some observations of a comet. The two men immediately became good friends but he was astounded to learn that Newton was not going to publish his monumental *Principia Mathematica*. Halley agreed to finance the publishing out of his own pocket, corrected the proofs, and arranged for distribution. In an address at Oxford several years later, a fellow professor said of Halley:

"…but for Halley the Principia would never have existed…He paid all the expenses, corrected the proofs, he laid aside his own work in order to press forward to the utmost the printing. All his letters show the most intense devotion to the work."

By 1691 Halley was looking for a teaching position at Oxford. Flamsteed, in the meantime, had become highly jealous of Halley and staunchly opposed his teaching appointment. Flamsteed also was at odds with Isaac Newton and, upon learning of his friendship with Halley, managed to stop Halley's appointment altogether.

In 1695 Halley was deeply involved studying comets and published his revolutionary theory that comets have predictable orbits, naming one special comet in particular. The comet of 1682, he declared, was the same as the comets of 1531 and 1607. He later identified it also as the comets of 1305, 1380 and 1456. Newton provided some much-needed mathematical support for Edmund's continuing work. Then Halley predicted the same comet would return in 76 years - in December 1758. Unfortunately he died in January of 1742 before he could see the visible proof of his theory. Halley later had succeeded Flamsteed to become England's Astronomer Royal.

In 1986 I stood out in a field at a little town in the middle of Missouri, camera and binoculars in hand, trying to get a decent sighting of Halley's comet without the distraction of city lights. The fuzzy object looked almost like a dim cotton ball but it was impossible to get a picture of it. After all the work I had put in preparing for the comet's appearance, it was highly disappointing. No matter, however, in a short time I would have the best of all possible views – close-up pictures of the comet's nucleus.

Halley's Comet Return Appearances

First Sighting by hinese				
	240	684	1607	
	164	760	1682	Halley
	87	837	1759	
BC	12	912	1835	⎫ Mark Twain
AD	66	989	1910	⎭ Lifespan
	141	1066 Hastings	1986	
	218	1145	2062	
	295	1222		
	374	1301 Giotto		
	451	1378		
	530	1456		
	607	1531	**Chart by T.Becker**	

At the 18th General Assembly of the International Astronomical Union meeting in Patras, Greece on August 26, 1982 the IAU endorsed by resolution the International Halley Watch as the worldwide receiving agency for Comet Halley observations. This cornerstone agency became the official coordinator agent for all professional and amateur data to be developed at the time of Halley's Comet return in 1986. The European Space Agency, working closely with the Academy of Sciences of the USSR, had a plan by written agreement to send a highly sophisticated spacecraft to intercept Halley's Comet and obtain images of its nucleus. Such a feat had never been done before but if it were really possible to achieve, it would forever change the study of astronomy. As it turned out, Europe and the USSR were part of an armada of six spacecraft sent to rendezvous with the comet.

Two Soviet spacecraft named Vega 1 and Vega 2 would act as the Pathfinder craft feeding mathematical and physical data to ESA in order to confirm and help guide ESAs Giotto spacecraft to the nucleus of the comet. The Giotto craft would encounter Halley and slide past the comet at (hopefully) a minimal distance of about 600 to 400 miles all the while operating its cameras to image the comet's physical properties. The mission of the Giotto craft was tricky – much like standing in Kansas City and hitting a golf ball precisely between the legs of the St. Louis Arch.

Spacecraft	Country	Launch
Giotto	European Space Agency	July 1985
Vega 1	Soviet Union	Dec 1984
Vega 2	Soviet Union	Dec 1984
Suisei	Japan	Aug 1985
Sakigaki	Japan	Jan 1985
Internat'l Cometary Explorer*	United States	Aug 1978

*Rerouted from its previously assigned mission

The Florentine master Giotto di Bondone painted a scene "Adoration of the Magi" in about 1303-1304. He used Comet Halley as a representation of the Star of Bethlehem appearing over the manger scene in the painting. Halley's Comet also appears on the famous Bayeux Tapestry above the figure of King Harold as a scene from the Battle of Hastings in 1066 which is considered the birth date of the British Empire. King Harold saw a comet just before the Battle and believed it foretold his doom. He was right; he was killed in the midst of the Battle.

Throughout history, a comet was a harbinger of doom or of great catastrophe. In 648AD, a comet hung over Nuremberg, Germany - the city had three months of rain, a deadly visitation from the plague, and eclipses of the Sun and the Moon. A comet appeared just before Hernan Cortes hanged the Aztec Chief Moctezuma in 1531. There was a rash of suicides in America as Halley's Comet appeared in 1910 because people learned the comet's tail contained large amounts of cyanide gas and the Earth was predicted to pass through the comet's tail. Gas Masks were sold on street corners across the nation along with souvenirs and food vendors' wares.

After a flawless launch by a French Ariane 1 rocket at the ESA launch site in Kouru, Guiana (South America) on July 2, Giotto (above) arrived at the comet on time, on course and came within 596 kilometers (about 370 miles) of the comet nucleus with instruments working flawlessly. Photo courtesy European Space Agency.

The nucleus is peanut- or potato-shaped with a mass of about 100 billion tons. The surface is irregular-shaped and the nucleus is filled with pits. The nucleus is composed of ice, stony compositions, and heat-resistant lightweight material. The jets emitted from only about 10 percent of the comet's surface leaving 90 percent inactive. Although the comet was formed in the outer reaches of the solar nebula, it has a surprisingly low abundance of carbon and nitrogen content.

Gas is emitted from the nucleus at about 20 tons per second, made up of 80 percent water, 10 percent carbon monoxide and 10 percent other gases. Dust is emitted at between 3 and 10 tons per second but only when the comet is near the Sun, suggesting an average loss of about 100 million tons although the comet has a total mass of about 100 billion tons.

(This report is paraphrased from Balsiger, H. et al. "A Close Look At Halley's Comet", Scientific American magazine, Sept. 1988 two years after encounter.)

The project confirms the theory of Harvard astronomer Dr. Fred Whipple who calls comets "dirty snowballs." The emitted gases called "jets" are often strong enough to push a comet off its course slightly which may make Halley off a day or so in its arrival near Earth. Gases absorb invisible ultraviolet sunlight then reflect it as visible light. The comet is heated by the Sun and begins to melt creating gas jets and a long tail. Comets are made up of rock, dust, ice, carbon, nitrogen and hydrogen.

By the time Giotto almost reached the comet, I had long been in contact with the European Space Agency's Washington D.C. office where Ian Pryke, ESAs Head representative in the United States at the time, had been supplying me with various kinds of information. Seated at home in my study, clipped articles and photographs littered my desk and the floor around it including a superb set of color slides prepared by ESA for use by scientists and organizations.

The task looming ahead – giving lectures of the Giotto mission and its discoveries to students of all ages – seemed an impossible chore because of the complexity of the technology and because of the intricacies of the mission itself. The closer to the actual mission the more time had to be devoted to it. Did the launch proceed as planned? Was Giotto on course? Was the data coming in correctly? Were we getting prime data at comet encounter? Dozens of questions had to be investigated and answered and the data had to be archived and accounted for. It all finally came together slowly in bits and pieces and the students were delighted.

All the participating spacecraft met with success. Giotto was able to home in on the Vega 1 and 2 guidelines straight to the heart of the coma. The Russians were faced with some problems of imagery refinements which they were able ultimately to do without complications. The two Japanese spacecraft operated flawlessly sending back data that correlated well with information from other craft and confirmed data all around.

Fourteen seconds before actual encounter, Giotto was hit by a sudden jolt from a tiny but fast-moving dust particle that was big enough to disable several instruments and temporarily push the spacecraft ever so slightly from its intended path. In the long run, however, the difference was so small that excellent images were still possible. Giotto's dual bumper shield was designed to ward off most particles – this one, however, got through the net.

The return of Halley's Comet invariably brings to peoples' minds the question of where comets come from. According to current theory, for our Solar System comets originate in a complex location called the Oort Cloud. The theory was first put forward by an Estonian astronomer named Ernst Opik in 1932 and often is referred to as the Opik-Oort cloud. The theory was revitalized in 1950 by Dutch astronomer Jan Oort who questioned some of the basic ideas about the Cloud.

Our Solar System does not end with Pluto, the fartherest of our planets. The Oort Cloud is situated at the very edge of our Solar System, out beyond Pluto, and extends more than 3,000 times farther. The oval-shaped Cloud is made up of comets left over from the formation of the solar system, a vast and cold region filled with comets that come and go according to their own private schedules. There may be trillions of typical long-period comets (such as Halley) in residence; occasionally one of them finds its way into our Solar System and then begins its elongated journey in a familiar periodic cycle.

No one has ever seen the Oort Cloud, but then no one has ever seen an atom either. There are trillions of pieces of cometary debris in the Oort Cloud, suspected by Tycho Brahe who studied the comet of 1577. In 1705, Edmund Halley identified 24 comets in his catalogue and argued that their orbits might be very long ellipses around the Sun. Today it is believed that *elliptical comet orbits* are on a predictable "come and go" pathway since they have been captured by the Sun. Their movements also are influenced by the planets, hence their rather varied timetables. There are other comets, in *hyperbolic orbits*, whose paths take them back to interstellar space where they began. These comets are very weakly influenced by the Sun and a rogue passing star can exert a strong influence on them.

The Opik-Oort Cloud is an unconfirmed theory. Astronomers have been studying it for decades but its complexity, together with the fact it is generally inaccessible by today's equipment and technologies, makes the Oort Cloud a very difficult object to pin down. We still rely, to a surprising extent, on the Renaissance observations and calculations of Tycho Brahe, Johannes Kepler and Edmund Halley. Sometime in the future perhaps a new theoretical level or a more definitive kind of technology will allow us to begin breaking the code of the Oort Cloud.

All over the world, the return of 1986 Halley's Comet was celebrated by a global greeting that extended from Chicago to Australia to Tokyo and London. The party atmosphere was a reminder that Halley was a human comet, its 76 year return period about the same span of a human lifetime, its return always cause for celebration and reminiscing. Most people have the opportunity to see the comet only once in a lifetime; only a privileged few have a chance to see it twice.

By the light of a late night lamp spreading a glow of warmth across my desk, I wondered what had been accomplished by these difficult missions into the starry heavens in search of one legendary object. Surely, I felt, Giotto had spawned a new technology and paved the way for the study of future cometary appearances that had never been possible before now. In the human quest for more knowledge about the universe, the 1986 event proved it could be done. In the midst of the many diagrams and mathematical equations that led up to the launch of Giotto came new ideas and perspectives about how to aim and target the nucleus of a comet. Here were two objects in simultaneous motion at different speeds and from different directions that could meet and join at one invisible point in the universe.

Two men scratching their names on a piece of paper – Raoul Sagdeev for the Soviet Union and R.V. Bonner for the European Space Agency – welcomed this great moment for science that transcended the Cold War and the clash of political ideals. Comet Halley provided an opportunity for scientists to get to know each other better and to go beyond Earth-bound political differences.

Within a few days a package mailed from the European Space Agency reached my home, via the Max Planck Institute in Germany – two exquisite images of the nucleus of Halley's Comet. In glowing color and crystal clear digital resolution, they were a stunning statement of the epitome of space technology in the latter half of the 1980s. I had to remind myself of what it had taken for the images to get to me. After an eight month journey across 434 million miles of hostile space, traveling at roughly a speed of 242 miles per second, a little spacecraft carrying cameras captured images of the nucleus of a comet in such detail that Halley's landscape could be discerned with startling clarity. How impossible it all seemed, yet it was done with a confidence that seemed almost off-handed in its execution. Because of the willingness of men and women (whom I had never met) of good will to provide me with information, I was able to describe the entire scenario of Halley's return to eager young people who, like me, were ever so grateful to ride the coattails of hard-working scientists.

Halley will be back in the 2061-62 time period and the excitement will begin all over again. Who knows what new technologies will be operating in that far-off time? What methods will we use to celebrate its return next time?

8 RUSSIAN SECRETS

"Do not retreat one inch; die at your guns, but save our glorious homeland."

Premier Josef Stalin, 1942
The Battle Of Stalingrad

Russia in the 20th century was always a land of turmoil. The causes of more recent official national behavior can be found in the multiplicity of somber events over several centuries of repeated invasions. From the beginning when it was nothing but a struggling little village of peasants huddled at Kiev, the history of Russia has been a saga of repeated wars, the failure of leaders without vision, and the enormous expanse of its own landscape. Even as recently as Napoleon's invasion in 1812, when the French armies ravaged much of Russia right up to and inside the gates of the Kremlin in Moscow, Russia never recovered from its entanglement in one war after another.

The years 1900 – 1950 form the crucible in which modern Soviet policy was forged in steel and blood and violence. The loss of the Russo-Japanese War of 1905, and the unfortunate mistake committed by the Czar's troops on Bloody Sunday that year started the 20th century off on a terrifying foreboding of larger calamities to come. No sooner had World War I begun on European battlefields in 1914 than the Russian Revolution erupted in 1917 at home. Nikolai Lenin's return from exile to the Finland Station set into motion cataclysmic forces that succeeded in creating an avalanche of still more immediate calamities.

The insecure communist revolution provoked a civil war in the 1920s that brought about untold suffering throughout the land. Lenin's death in 1924 allowed the rise to power of Joseph Stalin, an equally callous and brutal dictator whose Five-Year plan in the midst of severe economic depression never succeeded with the desired results. His ruthless rule became no more than an extension of the previous nightmares of political, economic and agricultural savagery. Several military and political purges of cultural leaders with assassinations fed Stalin's dormant paranoia, clearing his path of possible obstacles to his success. As World War II began, Russia once again was caught squarely between German advances across Europe in the west and Japanese conquests in China to the south.

Stalin's non-aggression treaty with Adolph Hitler was a desperate bid to protect Russia's borders. The Soviets immediately moved into Latvia, Lithuania and Estonia to create a buffer zone against rampaging German armies. Hitler's betrayal of the treaty in 1942 was a Soviet calamity without parallel as the Soviets lay naked in front of Hitler's blitzkrieg. Only by sheer force of will were the Soviets able to turn back the German armies, often in hand-to-hand combat in a battle line that scarcely wavered more than a few feet, to save their own country from certain devastation.

As World War II came to an end and the conferences at Yalta and Potsdam mapped out plans for European reconstruction, Stalin once again was looking for methods to protect Soviet borders. Soviet trust in other nations had been broken for all time to come. For the Soviets, self-protection became an ingrained and all-pervasive national policy as did refusals to disclose Soviet intentions and a break-neck defense that kept a wall of silence between Stalinist politics and the democratic Western nations.

The partition of Germany and the communist take-over of Eastern Europe offered the surest possible wall between the Soviet Union and the sub-continent of Western Europe. Filled with continued fears of betrayal, duped by the very allies alongside whom they had fought, a ruptured Soviet Union emerged from World War II broken and shattered from years of struggle, hardship and misery. Looking forward to the remainder of the 20th century, Stalin was ever mindful of his country's future stance in world affairs. "Never again, comrades, will we allow any nation to threaten us," he shouted. "Promise it to your children and to your grandchildren for all time to come." So it was promised over the graves of 20 million Russians who had perished in the war; so it was done.

Stalin's paranoia, now almost out of control, triggered a response among western observers initiating a series of agencies that must have astonished not only Stalin himself but the rest of the world as well. The United States, sensing a growing global prelude to a third world war, began a massive protective build-up of weapons and weapon systems which were pretty well in place even before Sputnik was launched in October 1957.

Strategic Air Command (SAC) 1946

At the end of World War II, by March of 1946, the US Army Air Force was being split away from an Army function and became the United States Air Force in mid-September of the following year. Immediately, General George Kenny was named in charge of a special Air Force group named the Strategic Air Command with the responsibility of building an American global protection and attack capability. Starting with only a handful of bombers and fighter planes, the USAF became the most awesome global air power umbrella in history. SAC's original directive included four main areas of activity

1) long-range offensive operations in any part of the world
2) maximum-range reconnaissance over land and sea
3) provide combat units capable of intense and sustained combat operations employing the latest and most advanced weapons
4) undertake special missions.

In short, the Strategic Air Command was to become a "do-it-all-and-get-it-done" global defense and strike initiative as a response to a Soviet threat and a prelude to World War III. General Curtis LeMay took charge of SAC in 1948 and within a year he was conducting training exercises for a totally unprepared air arm.

North Atlantic Treaty Organization (NATO) 1949

NATO is a military alliance established by treaty on April 4, 1949 with headquarters in Brussels, Belgium. It is a collective and mutual defense organization originally of 12 nations, with two more added in 1952. NATO commands an active *NATO Response Force* whose efforts are limited to North American and North Atlantic countries. Today the membership stands at 26; military forces of separate nations are decided by individual NATO/Nation agreements. When West Germany was admitted to the organization, the Soviets responded by creating the **Warsaw Pact in 1955** composed of the USSR, its allies and its group of East European nations.

Arctic Distant Early Warning Line (DEW) 1957

To prepare for armed aggression coming over the North Polar Cap (from Russia), it was necessary to install a warning system north of the Arctic Circle. The system was named the Distant Early Warning Line – a fence of radar stations across the top of the world. Begun on paper in 1952, it was a shared activity with Canada and a string of 63 radar and communications sites stretching 3000 miles from northwest Alaska to Baffin Island just east of Greenland. The Line was constructed by 25,000 workers in the worst possible environment of freezing weather and extremely desolate landscapes. In 1985 the name was changed to North Warning System and in July 1993 operations of the DEW Line were turned over to Canada with headquarters at Ottawa.

North American Aerospace Defense Command (NORAD) 1958

NORAD was founded in May 1958 as a joint effort by the United States and Canada as an aerospace warning and control agency. The organization's technical facility is inside Cheyenne Mountain near Colorado Springs, Colorado supported by an operations center at Peterson Air Force Base, Colorado also just outside Colorado Springs

No sooner had Sputnik been sent aloft than there was a scramble to find the Soviet launch sites. Central Intelligence Agency U-2 overflights had as a priority the search for and identification of the sites. For public relations purposes, the Soviets announced their site at Baikonur Cosmodrome in the Ukraine region. A thorough search and mapping project began that failed to find the site. The announcement was a ruse – the Russians weren't telling the truth as usual and locating the site was highly important since it would provide ongoing clues to Soviet space progress. The CIA eventually pinpointed not only the main site for manned launches but other launch sites at Plesetsk (unmanned launches) and Kapustin Yar. But the CIA had help. The breakthrough came from a highly unexpected source; a British school teacher and his students in the north of Britain.

At the Kettering Boys School in England in the 1970s and 1980s, Geoffrey Perry was purposely teaching classes in modern technology but they were forced to use outdated equipment. Often his students had to scour half of England to find original spare parts to build radio receivers and other kinds of electronic devices to match Soviet far-outdated and no-longer used equipment. As Soviet launches continued without let-up, Perry became enamored with the problem of finding the launch site. Using only mathematics and the laws of physics, he and his students tracked objects being launched, applied mathematics to their trajectories, orbital periods and altitudes, and discovered the Soviet manned launch site at Tyuratam. It was a monumental piece of detective work.

Perry and I taught at the British Space School first in 1989 at the Sevenoaks School in Kent and thereafter at Brunel University in Uxbridge. The electronic detective work involved was seemingly simple in its explanation but much more complicated in truth. They had the time factor – launch to orbit, Perry said. Then they were able to compute the trajectories and orbital time periods along with the speeds of the spacecraft and rockets, he continued with a twinkle in his eye. After determining those figures it was merely a matter of constantly confirming at launch after launch the recurring paths of the objects being sent into space. Finally they had it all pinned down.

Perry's students also were in the habit of recording sounds sent out by various Soviet satellites and spacecraft. At the time, every space-intensive country was playing music and speeches from space using an onboard music generator. Perry was kind enough to give me two such recordings. One was the *Communist Internationale* from a Soviet satellite, and the other a rather more intriguing recording from a communist Chinese satellite playing a movement from *The Yellow River Concerto* featuring the section *The East Is Red*. With a little digging I managed to turn up a Christmas greeting broadcast by President Dwight Eisenhower from an American SCORE satellite launched in December 1958. The broadcasts from space were very effective propaganda in the early days of space technology development.

Geoffrey Perry (left) working on an equipment installation project during the British Space School at Sevenoaks School in Kent, England.

Since Perry had located the launch site, the next logical step was to see what it looked like. That was a job for the CIA and their U2 aircraft. The problem was not that easy – it meant the use of U2 overflights *inside Soviet borders*. This also meant asking for an international incident if anything went wrong. Luckily nothing went amiss and in 1959 a CIA flight originating at Peshawar, Pakistan at 70,000 feet caught the Tyuratam manned program launch site dead on.

Additional overflights also were able to gather repeated launch site imagery of both Plesetsk and KapustinYar. President Eisenhower voiced his great concern even while he was signing approval for the flights. His fears became reality when Francis Gary Powers was shot down in 1960 by a Soviet Surface-to-Air Missile and was put on trial in a Soviet court for the entire world to see. Having been found guilty, Powers was sentenced to several years in prison but was traded with the USSR after only two years for a Soviet spy caught in the United States. The Soviets had been able to track the U2 flights but weren't able to catch up to them until they developed the SAM that shot down Powers.

The intelligence community is a worldwide brotherhood of allies that share spy photos and information. Seemingly out of nowhere in 1982 came a whisper of some Soviet clandestine activity in the Indian Ocean. The Soviets had a "fishing trawler" on station with high level sensitive receiving equipment – waiting for something to happen. It didn't take long; a parachute, an aerodynamic object with a tall cone-shaped covering protruding from its forward section and a shorter cone near the center of the craft, a slow descent and a splashdown. NORAD had tracked it at launch and continued to track it now out of orbit. An alert went out to numerous western governments.

The Royal Australian Air Force dispatched a P3 Orion reconnaissance aircraft to the location to check on activities there. The Orion made several photographic passes around the trawler and discovered a Soviet craft of unknown purpose. The pictures sent to me by the Australian Ministry of Defence at Canberra seemed to offer several possibilities ranging from a prototype model of a manned spacecraft in the design research stages (the craft was felt to be a one-fourth scale design) to a model of an attack spacecraft that could be used to shoot down satellites or even a Space Shuttle or, at the extreme, to disable a Space Station. It also could be a reduced design shuttle-type vehicle to take cosmonauts to a Soviet Space Station and return cosmonauts to Earth, or perhaps just a re-supply ship. In any event, the craft signaled a growing capability of considerable breadth to the Soviet space program of which western observers were not completely aware.

The Australian photographs show a winged spacecraft with a cone sticking out of the forward fuselage, probably to protect communications electronics. Full-suited scuba divers retrieved the craft and prepared it for hoisting aboard the Soviet trawler. The craft was being fished out of the water by a winch and placed on the open deck where it was made ready for storage in the ship's cargo hold. Since the Royal Australian Air Force plane arrived on the scene in record time, its cameras recorded numerous different views of the retrieval process right from the start.

Soviet "spaceplane" one-fourth size test model lifted from the Indian Ocean, photographed by an Australian reconnaissance aircraft. Photo courtesy Ministry of Defence, Canberra and the Royal Australian Air Force.

In other photos, personnel on board the ship wore gas masks and were covered head to foot by rubber suits, probably as protection against possible radiation and other types of unknown space chemistry. Since Soviet personnel were on deck to maintain the craft, they offered an obvious size comparison. Compared to the dimensions of a human, the craft was determined to be a model being tested for research purposes and not a full sized finished spacecraft. This first event took place on June 3 1982, the craft being designated by US Department of Defense analysts as Cosmos 1374. In all, there were eight separate trials of the model vehicles supposedly to make various internal and external design changes, and included Cosmos 1445 in March 1983, Cosmos 1517 in December 1983, Cosmos 1614 in December 1984, and so on.

It was a widely known fact that the Soviets all along had been planning for an expanded manned space program. The main problems for all space faring nations in the 1980s and 1990s revolved around two specific unresolved technological capabilities. The first problem is what aerospace engineers called *single stage to orbit*. Spacecraft carrying humans into space relied on three-stage rockets resulting in higher costs and heavier rockets. The Apollo-Moon rockets (Saturn 5) were three-stage rockets that required massive amounts of fuel and the building of three separate rocket sections for adequate thrust. It became imperative to be able to get from the launch pad into space using only one rocket section.

The second big problem, and a critical one at that, is *an extended presence in space*. If any nation is to conquer space for travel, research or commerce, it is necessary to live in space for long periods of time or else humans would forever be just cosmic visitors. And if we are really serious about establishing research bases or human colonies on the Moon or on Mars, we need to stay in space for many months or years. Finding people to go to these frontier outposts would never be difficult, but devising the means for them to stay there and survive will be a hard task for some of the best minds on the planet to research.

There have always been plans and designs for spacecraft to remain in Earth orbit for long periods of time. America created the first of these craft as the Space Shuttle. It also was a good bet the Soviets were designing a similar craft but Soviet silence and refusal to publicly disclose long term plans made it difficult to assess progress. In the Cold War environment and America's need to continue to build international respect, it was important for America to continue in first place. But who was ahead in this new race? Was there really a "space race" in the 1980s and 1990s? Determining these answers became a national priority for both countries.

The Soviet penchant for modification of existing spacecraft designs instead of designing completely new vehicles was a continuing policy in the last three decades of the 20th century. Having found a basic design that worked well, the Soviets were unable to abandon it in favor of new and untried systems probably in fear of losing time required for new flight testing. The Vostok (Yuri Gagarin's spacecraft) was modified to become the Voskhod three-man spacecraft which was re-modified to become the Soyuz manned spacecraft for ferrying cosmonauts and supplies back and forth to the Salyut 1st generation space station. Even when the MIR space station was put into orbit, the Soyuz craft continued to be the Soviet space workhorse.

The overall Soviet political and strategic plan in the 1980s and 1990s relied upon three vehicles and two required achievements. The **Soviet Triad** was made up of an orbiting and continually supplied manned space station (**MIR**), a heavy lift launch rocket (**Energiya**), and a recoverable manned spacecraft (**Buran**), all of which would lead to a *permanent human presence in space* as well as a *strategic leadership capability* in case of a space-based global war. Neither the Americans nor the Soviets had yet conquered the problem of a single-stage-to-orbit manned vehicle although both countries had advanced designs for setting up space stations.

Sketch of Soviet spacecraft

Early in 1971, the Soviets began a long series of projects. On April 19, 1971 the Soviets launched its first space station – a small craft named Salyut 1. Its crew remained on board for 23 days when an accident caused the atmosphere to escape and the crew perished. The station carried a gamma ray telescope and a spectrographic telescope. In that 23 days, the crew conducted Earth observation, studied the biological effects of micro gravity on plant growth and nutrition, and the long-term effects of microgravity on human organisms.

A greatly modified Salyut 2 station was launched in June 1974 but failed to achieve orbital stability and it broke apart before a crew could be put on board. Salyut 3, modified once again, sported larger solar panels, upgraded life support and power systems, and a more comfortable interior. A crew remained on board for 14 days and conducted some 400 experiments but, when a relief crew was unable to dock with the station, it was abandoned and the crew returned to Earth. Salyut 4 launched in December 1974 rotated its crews while the space station remained aloft for 770 days before it was brought down to Earth. Salyut 5 launched in June 1976 remained in orbit for 412 days rotating two different crews.

Salyut 6 was sent up in September 1977 as a fully modified craft and stayed in orbit for 676 days thanks to most of its onboard systems being heavily modified. Research was conducted in many different scientific categories including astronomy, Earth observation, materials processing of glass, super-conductors and ion crystals, and biomedical experiments. Salyut 6 was in orbit four years and 10 months with a crew accumulation of 676 days. America had slipped behind the Russians again.

Salyut 7 launched in April 1982 was the most up-to-date and fully automated space station in existence. The first crew conducted 300 experiments on request from 500 Soviet scientific institutions and science centers. Celebrities among crew members included the first French astronaut Jean-Loup Chretien and the world's second woman in space Svetlana Savitskaya. A new modular transport craft, Cosmos 1443, successfully docked with the station adding to the available on-board amount of space.

The Soviets characteristically never pre-announced their major launches; free-world intelligence operatives and monitoring organizations were hard pressed to keep up with Soviet space activities in the 1970s and 1980s. The American-led global intelligence gathering community was supported in part by a Soviet underground spy network that fed data and photographs to the CIA at regular intervals. A number of underground Soviet citizens paid with their lives to deliver high risk intelligence data. One Soviet citizen I learned about made sketches of spacecraft on the back of an envelope he secretly delivered to a CIA operative. He was later caught by the KGB and executed. An American space expert at the time once quipped that there probably were more spies in Moscow than Soviet citizens.

Soviet cooperation with the free world was limited mostly to specific projects and on a theoretical level although there were some joint space missions such as the Apollo-Soyuz Test Project (ASTP) better known as the "handshake in space" (see Chapter 1). SARSAT (Search And Rescue satellite), a joint Soviet, American, French, Canadian and British project was a search and rescue satellite team using distress signals from portable Emergency Locator Transmitters on board aircraft and ships that broadcast an internationally recognized distress signal.

The Soviets drew numerous crew members from India, France and Japan as well as from Soviet Bloc nations - Poland, Czechoslovakia, Bulgaria, Hungary, Vietnam, Cuba, Mongolia and Romania. Luckily all these space flights were textbook forays mostly for public relations benefits since the foreign crew members for the most part were just passengers.

The gap between a highly successful Soviet space effort and grounded western space vehicles became wider in 1986 than at any other time since 1957. In fact the Soviet Union was the only spacefaring country in 1986 as the Soviets began an extremely intensive effort in that decade. The Soviet grand finale came in the 1980s with the launch of the Space Triad. The first section of the MIR space station, the first block of what was intended to be a mammoth orbiting space station-city, took place in February 1986 followed by a continuous fleet of supply vehicles.

In May 1987, the *Energiya* heavy lift booster was launched for the first time. In November 1988, the launcher carried the first Soviet *Buran* ("snowstorm") space shuttle on a flawless electronic-guided flight remarkable for two reasons. It was the first Soviet pre-announced space flight, and both launcher and shuttle craft were still in their infancies. As the Soviet attitude toward the free world mellowed considerably, all these activities were announced beforehand. The Soviets had achieved their goals.

**Art concept of the launch complex at Tyuratam showing the
Energiya heavy-lift launch vehicle and the Buran shuttle-type
space vehicle. Artwork courtesy U.S.Dept of Defense.**

The Russians worked on many other offensive and defensive systems in secret. From the German scientists conscripted to build the new Soviet Union came details of the weapons invented by the engineers at Peenemunde. The Department of Defense also sent me in the late 1980s photos of artwork of submarine-launched missiles, a railroad-launched missile system copied from the Nazis V2 *meillerwagen* concept, huge antennas and electronic listening structures, and massive radar defenses. The art concepts were created from many different kinds of intelligence gathered by hundreds of unnamed operatives working quietly around the world. From all these inventions and systems it was obvious the Soviets maintained a continuous ready-state preparation for World War III. Premier Stalin's paranoia was absorbed by the Politburo and the Central Committee despite *Glasnost* and *Perestroika* and is still alive and well in the new Soviet Union.

9 HURRICANES ON OUR DOORSTEP

"Rock-a-bye baby in the tree top,
When the wind blows the cradle will rock,
If the bough breaks the cradle will fall,
And down will come baby, cradle and all."

Olde English Rhyme

My introduction to hurricanes came aboard a US Navy rocket launcher in the Caribbean Sea in the mid-1950s when I was only 20 years old. The ship was headed for homeport at Norfolk, Virginia when we became entangled in the outskirts of what proved to be a major category storm – and the most frightening experience of my life. A later attempt to write down the experience as best I could remember in the middle of a life-threatening situation describes the event.

"The storm's first waves breaking over the ship's bow were estimated at eight to ten feet. With a roar, they splashed across the foredeck washing the rocket mounts and the 5inch .38 gun mount of the Navy LSMR rocket launcher. The flat-bottomed ship heaved upward, digging the fantail into the black sea and pointing the bow skyward. Then, with a shudder, the ship's bow dipped and fell back down into the angry water in time for another wave across the bow. The up and down motion in the heavy sea continued for some six hours then worsened as, nearly out of control, we came even closer to the hurricane's eye. By following morning, bow waves were almost twenty feet high, lashing the decks and smashing against the ship's bridge with a deafening, terrifying roar. Despite the helmsman's strength and skill, the ship nevertheless suddenly

skewed sideways with a sickening spin and slipped down into a deep sea trough.

Tied with rope to a chair at the radar set, I could feel my stomach getting queasy and nauseated. After some begging, I coerced a shipmate into untying me. It was a mistake I later regretted. Holding onto handrails and the edges of furniture and equipment, and dodging chart books and navigational aids, I was able to reach the hatch to the outside deck, cracked it open and tried to peek out. In a split second, the door was ripped from my hands and flung wide open as the flat-bottomed Navy rocket-launcher fell downward again into the black, watery coffin of an Atlantic trough. Cold rain stinging my face and hands, I stared in abject horror at the immense fury of the angry sea spitting and seething with gray-black, endless water. Turning to look straight upward above the ship, all I could see was the threatening, swirling ocean. Obviously our ship was nearly on its side. Then the ship rolled with an enormous heave and a low sounding growl, lifted swiftly almost entirely out of the water in a mighty upward surge. As I looked far downward over the side of the ship, all I could see was the sickening, near-black pallor of the angry sky. Then the ship suddenly came crashing downward only to repeat the cycle all over again. With a panic I had never known before in my young life, I murmured aloud, "Oh, God, I don't want to die this way!"

During the hurricane season each year in the North Atlantic Ocean - *June 1 to December 1* - meteorologists and national hurricane monitoring stations all over America track the season's hurricane activity. North Atlantic hurricanes especially are of interest to America because they pose the greatest threat to life and property on the East Coast and in the Gulf of Mexico. Pacific hurricanes also are tracked under the names of *cyclones* and *typhoons*, but they travel away from Pacific Coast states because of the Earth's rotation and threaten only the American Hawaiian Islands. During their cycles, cyclones can cross the entire Pacific Ocean and attack Japan and China.

Records are kept for each season going back to the late 1800s. Researchers use these records to determine important patterns on a century-long basis, furnishing statistics that are applied in a number of different ways; heavy or light seasons, patterns of the number of various Categories of hurricanes, length of life, and so on.

Hurricanes are like fingerprints; no two of them are alike. However, researchers who work with and study data about hurricanes have disclosed that hurricanes, like humans, pass through a *definite life cycle*. The cycle begins with a low barometric tropical depression and then moves from one life stage to another until it finally dies out. The entire cycle looks like this:

Wave
tropical depression
tropical storm
Category 1
Category 2
Category 3 (over water)
Category 4
Category 5
Category 5+
tropical storm (after landfall)
tropical depression

Hurricanes need two essential ingredients to function: *heat* and *moisture*. If one of these ingredients is absent, the hurricane cannot sustain itself. In the North Atlantic, both these features usually are available in abundance. Atlantic hurricanes form up and travel near the equator (furnishing heat), or push off the coast of western Africa, or form up in the Caribbean or Gulf of Mexico immediately encountering ocean water (furnishing moisture).

Even a high-category storm can die out if it passes over a landmass. The path they travel often is often referred to as the "North Atlantic Hurricane Highway." The path a hurricane takes is pretty much in a straight line across the ocean, falling onto whatever is in its path or moving in a jagged course depending on a variety of storm *Steering Devices*. The devices usually include combinations of effects so that actually predicting where a hurricane will go can be very tricky at best and requires real-time knowledge of meteorology at the time the hurricane is active. Following are some but not all of these devices:

Gulf Stream (*North Atlantic Drift*) activity
Jet Stream movement
Upper and Lower Level Winds
Ocean Currents and directions
Sea Surface Temperatures
Wind Directions and Forces from Landward
Hurricane Forward Velocity
Time of the Year/Month
Sun activity such as a Coronal Mass Ejection
Internal Strength of the Hurricane

Hurricanes begin life inhabiting the tropical zone of the planet. Heat from the Sun beats down on the ocean's surface heating up the already warm water and setting in motion a transformation of moisture from cool to warm. At the same time, the atmosphere and water droplets together begin a slow twisting dance that grows into a swirling mass. As the mass develops it becomes what meteorologists call a *wave*; the barometric pressure inside the mass then begins to drop slightly. The higher the heat, the more active the mass becomes as it moves through the hurricane's cycle.

The Earth's rotation eastward carries the depression westward and provides the mass with an endless supply of warm moisture. The mass continues to grow into the next stage of development called a *tropical depression (TD)*, and then grows into a *tropical storm (TS)* as the swirling mass picks up speed. Once the mass reaches TS status, it takes on new meaning and a new threat to life and property as it prepares to enter the realm of "hurricane" status at about 74/75 miles per hour (Category 1).

The swirling mass is measured by two different speeds or velocities during its journey. One speed is its *forward velocity* over the water which sometimes can reach as high as 55 to 60 miles per hour. The second speed is its *internal velocity*, or rate of spin. The internal velocity is the one indicated on the Saffir-Simpson scale and can reach up to 160 to 165 miles per hour with temporary gusts up to 225 miles per hour. Both speeds are reported when recording or tracking a hurricane.

Hurricanes develop across the tropical zone where the Sun shines directly perpendicular to the Earth and creates the hottest region of the planet. As winter turns to spring and then to summer, the region continually gets hotter and hotter. The water, despite its enormous area and volume, also grows warmer as the sea develops heated water to feed the hurricane during its journey. The birthplace of North Atlantic hurricanes changes during the season doing a sort of shift over the long six months from June 1 to December 1.

In June and July, hurricanes are born in and around the *Gulf of Mexico*. By mid-July, the birthplace begins to move eastward out *into the Caribbean Sea* where the sea surface is warming more and more. By August, the birthplace is well into the middle of the Atlantic Ocean, and by September the birthplace has shifted dramatically all the way to the West Coast of Africa where it is acted upon by various physical forces at work on that continent. Then, as the season begins to die down, the birthplaces shift back toward the Gulf of Mexico in October and November as sea surface temperatures begin to drop below 80 degrees F.

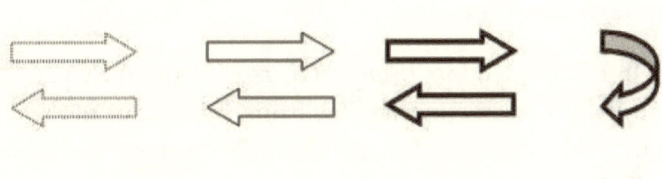

Sketch of arrows

From this point onward, the baby hurricane simply keeps growing through several stages of life following a measurement scale provided by a Miami engineer named Herbert Saffir in 1971. Sometime later Robert Simpson, once a Director of the National Hurricane Center, added Storm Surge and Barometric Pressure finally creating the **Saffir-Simpson Hurricane Scal**e we use today to describe a *hurricane's potential for destruction.*

Saffir-Simpson Hurricane Scale

Category	Wind	Barometric Pressure	Storm Surge	Potential Damage
1	74 to 95mph	More Than 28.91 inches	4 to 5 feet	Minimal
2	96 to 110	28.50-28.91	6 to 8 feet	Moderate
3	111 to 130	27.91 to 28.47	9 to 13 feet	Extensive
4	131 to 155	27.17 to 27.88	13 to 18 feet	Extreme
5	More Than 155	Less Than 27.17	More Than 18 feet	Catastrophic

Chart of Hurricane Scale

North Atlantic hurricanes *spin counter-clockwise* and, if they aren't affected by too many steering devices all at the same time, usually travel in a reasonably predictable straight line. Again this isn't always true, especially if the hurricane encounters the Gulf Stream that can carry it northward. Too, hurricanes seldom if ever turn around and backtrack (there are some exceptions), and North Atlantic hurricanes usually never dip or travel south of the equator. Below the equator hurricanes *spin clockwise*.

On Earth, the *Heat Transport System* is from the equator toward the North Pole. Ocean currents flow from the equator to the poles and back again. The tropical waters are heated from above and cooled from below, creating equator-to-pole transport of heat into the atmosphere. The Gulf Stream originates in the Caribbean and flows northward paralleling the U.S. East Coast. These Planetary Heat Transport Systems are very important in the life of the hurricane because heat is one of the two ingredients that feed hurricanes and keep them alive.

As hurricanes grow through the developing portion of their life cycle, they form an empty area in the middle analysts term the *hurricane eye* that meteorologists use as a major identification of and clue to the strength of a hurricane. As hurricanes strengthen, the eye's diameter in miles grows smaller and smaller. An eye with a diameter in the teens, for example 13 or 14 miles, isn't nearly as strong as the compact eye of 6 or 7 miles. Hurricanes with eye diameters below 10 miles are capable of great destructive power.

If the hurricane's forward velocity is great enough, the storm will drive across a piece of land and reach water again only to reassert itself and regain the power it temporarily lost. This behavior is seen again and again in storms that cut across the Florida Peninsula or which begin in the Gulf of Mexico and cut northeast across Georgia or the Carolinas and emerge in the Gulf Stream off the U.S. East coast. Although very dangerous, hurricane hunters fly around inside the eye to take measurements.

Few people realize how huge a mature hurricane is in sheer size. *Most major hurricanes* (F3 and above) measure about 800mi to1000mi total diameter, covering immense geographic areas and often about the size of an entire state. Because of their sizes, the storms inflict heavy damage to landmasses costing many millions of dollars. For example, a hurricane moving up the U.S. east coast shoreline from Florida to North Carolina can devastate several states because of high winds, heavy rainfall, flooding and electrical storms. Tornadoes frequently travel on the back edges of hurricanes; many hurricanes create their own violent weather systems.

Because of its spin, the hurricane pulls moisture inward toward its center and compacts the moisture into a tightly bound unit that forces the hurricane's eye to draw in upon itself. The diameter of the eye grows smaller and smaller, creating a mass of energy around the eye. The eye walls of the eye become filled with the energy, but the space *inside the eye is absolutely calm.* It is so calm, in fact, you could light a barbecue and cook hamburgers at ground level if you had the time. People caught in a hurricane mistake this calm as the end of the hurricane when in fact it is simply a lull in the storm and the other half of the hurricane is still on the way. The hurricane's eye is a tunnel of stored energy and water through which the hurricane breathes. Warm air moves upward and outward and then tumbles back down again. Very often, it is possible to see blue sky if you look up through the eye although it is highly dangerous to take the time to do so.

Statistically, the peak of North Atlantic hurricane activity comes about September 10. By then, the northern hemisphere begins to cool down. Occasionally a rogue hurricane will pop up in October or November but then the atmospheric, ocean and landmass temperatures are considerably cooled and the lack of heat all around just doesn't provide a sufficient environment to create a hurricane.

The hurricane eye is clearly visible in Hurricane Andrew '92, as well as the outside feeder bands of moisture being pulled in by the storm. Spin is counter-clockwise. Photo courtesy NASA.

One wall of the eye has passed, but the wall on the other side of the eye is still on its way and the hurricane's ferocity will start up again once the second side of the eyewall arrives.

The by-products of a hurricane can be devastating and cause the destruction to life and property most people associate with these storms. Torrential rains sometimes dump two inches of rain per hour, in turn causing flash floods and widespread flooding. Tornadoes can form in the perimeter of a hurricane with frightening displays of lightening and thunder. Wind gust from Category 3 or better storms can reach as high as 225 miles per hour. One of the most destructive of by-products is storm surge, especially in coastal areas sometimes reaching as high as 25-30 feet, pushed along in front of the hurricane at surprising speeds.

In 1995, both the National Climatic Data Center in Ashville, North Carolina and the National Hurricane Center at Miami helped me conduct a major research project for the 1995 season. Dr. William

Gray at Colorado State University (Boulder) had predicted a highly unusual and mammoth hurricane season – 16 Tropical Storms including 9 hurricanes with Florida the major strike zone. Dr. Gray has maintained an 85% success rate in his predictions – this year he was off only 1 additional tropical storm.

After studying hurricanes for fifteen years, I thought it was important to visit at least one site where hurricanes had hit the southeast fairly regularly to see how re-growth in the area had brought nature back to life. I chose Sullivan's Island, one of the barrier islands of the coast of South Carolina at the entrance to Charleston Harbor. It proved to be an ideal choice with a nice long beach and sufficient vegetation to show a comeback if indeed it did manage some good re-growth.

Working with the National Climatic Data Center (NCDC) in Ashville was a joy although coordinating computer information and satellite imagery was a bit tricky at times. Focusing on the entire range of hurricanes became hectic – getting up twice in the middle of the night to record data from the National Weather Service (NWS), making drawings of hurricane pathways, listening for velocities and storm surge numbers and other statistics became hectic after a while especially since the season started with a bang.

The first hurricane, a category 1 storm, hit the runway early on June 3. From that point onward, monitoring became a matter of keeping close watch on all methods of timely communication. Satellite imagery came in from the NCDC; storm statistics originated from the NWS and television's The Weather Channel; I had to be present to record the data. By mid-summer, I was driving breakneck down the East Coast from St. Louis to Ashville to Sullivan Island to Dothan, Alabama to Cape Canaveral to Miami and back again to St. Louis – never missing a data set. At times I was driving 85mph on the highway in order not to miss a data set.

One of the requirements for the project was to record one special hurricane in detail. I chose Hurricane Luis which became a Category 4 storm that tore up island real estate as it roared across the Caribbean. The visit to the National Climatic Data Center allowed me to photograph employees right at their desks as they tracked the hurricanes on their computers. Also, purely as a research specialist, I was happy to find that summer 1995 brought some record-breaking American city temperatures (shown below). That fact reinforced my belief that I had chosen the right year to do the project.

The technology developed to study hurricanes consists mostly of a fleet of special satellites designed to image moisture rather than land. All space-faring nations have their own weather satellites but the nations all work together to furnish data to each other. Other technologies include buoys, ships at sea that furnish eyewitness data, Doppler Radar for wide-area tracking, and of course the "Hurricane Hunter" fleet of aircraft that carry instruments and meteorologists. The aircraft fly down inside the eye of the storm.

City	Days Above 90 F On August 4
Philadelphia, PA	17
Raleigh, NC	23
Baltimore, MD	25
Washington, DC	25
Richmond, VA	26
Phoenix, AZ	(12 days 110 F or more)

	By August 18 – Temp 100 F or better
Columbia, SC	104
Athens, GA	103
Birmingham, AL	103
Charleston, SC	102
Atlanta, GA	101
Augusta, GA	100

Note: Omaha, NE – 40 days 90 F or higher by August

The hand-drawn chart below was part of an actual research project.

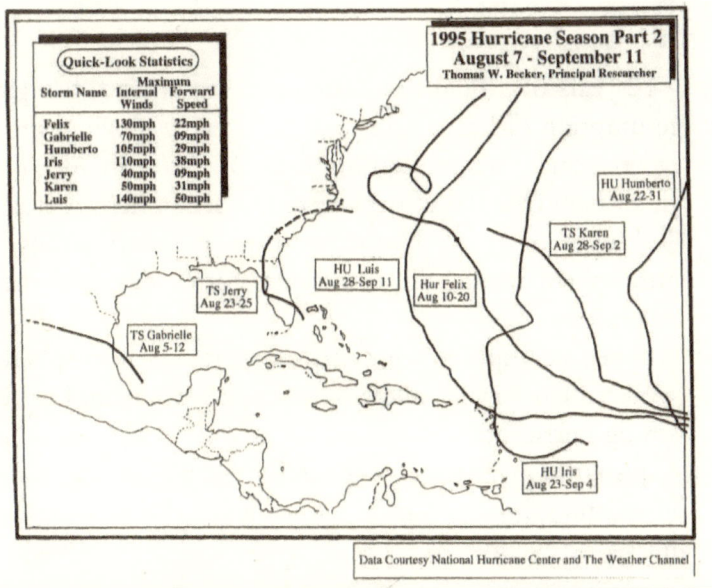

Sketch of map with hurricane tracks

On August 24 1995, Hurricanes Humberto, Iris and Tropical Storm Jerry
across the Atlantic Ocean were lined up like airplanes on a runway at Chicago's
O'Hara International Airport. Image courtesy NOAA/NESDIS-NCDC.

The reasons for these unusual 1995 temperatures were obvious and logical:

- 12 years of drought in Africa came to an end, bringing the resumption of heavy rains created by slow moving, moisture-laden clouds (waves) coming from *the Sahel* (border between the Sahara Desert and the vegetation of southern Africa);
- absence of an El Nino that usually creates strong upper atmosphere winds, did not carry storms out into the mid-Atlantic;
- upper atmosphere winds in the northern hemisphere were weaker than normal, allowing swift development of waves into strong hurricanes;
- springtime weather in the Caribbean with warm water and low atmospheric pressure encouraged the development of hurricane energy.

Hurricane strike zones, based on the 100-year archived record (roughly from 1889 to 1992) indicate the frequency of hurricane targets:

-Florida 55
-Texas 36
-Louisiana 25
-North Carolina 24

The worst storms during this same period, the Category 4 and 5 hurricanes, struck at:

-Florida 7
-Texas 6 (Galveston in 1900 killed 6,000 souls)
-Louisiana 4

Of 794 hurricanes 1899-1992, more than one-third (303 or 38%) crossed or passed adjacent to the United States mainland. Major hurricanes F3 and above have crossed the U.S. coast about twice every 3 years. In the same period again, 62 major hurricanes F3 or more have affected the United States. Only 10% (10 out of every 100) tropical disturbances ever reach hurricane status; of these, probably only 2 will strike the United States. As we move into the 21st century, all these figures will change slightly as the Global Warming/Climate Change scenario affects hurricane behaviors.

The "ocean watchers" of the world occasionally create a global focus on our oceans by sponsoring a worldwide project. In 1998, for example, the U.S. National Oceanic and Atmospheric Administration sponsored a special effort with the name *The Year Of The Ocean* in league with the U.S. Environmental Protection Agency and the National Fish and Wildlife Foundation. It was a year of intense ocean research designed to provide status reports that later could be acted upon by universities and specialized ocean monitoring agencies, and a chance for budgets to be made available to produce educational materials for elementary and secondary schools.

In 2000, Congress tasked the U.S. Commission on Ocean Policy (USCOP) to investigate and provide recommendations for a "coordinated and comprehensive national ocean policy." In July 2004, the USCOP published An Ocean Blueprint for the 21st Century with more than 200 recommendations, developed based on extensive hearings, written input, and public comment. In December 2004, upon consideration of these recommendations, the Bush Administration released the U.S. Ocean Action Plan (OAP). This broad plan proposed a fundamental restructuring of ocean governance, research, and management intended to "engender responsible use and stewardship of ocean and coastal resources for the benefit of all Americans."

President Bush proclaimed June 2007 the official "National Oceans Month" for the American public's calendar. The American national ocean policy was announced in 2004 and now serves as the United States policy and plan. On January 26, 2007, the ICOSRMI released the *National Ocean Research Priorities Plan and Implementation Strategy*, outlining 20 critical ocean research priorities for the United States for the next decade. Developed in response to the Ocean Action Plan, these priorities focus on the most compelling issues in key areas of interaction between society and the oceans.

10 MASTERWORKS OF TECHNOLOGY

Five Short Examples
Photos By The Author Except the Hubble Telescope

reative technology is not an American invention; it is a human skill developed by people all over the world. Technologically advanced cultures have more opportunities to increase and upgrade technologies, however, and may appear to be more advanced technologically because they can invest more financial resources to create a variety of technologically superior products and systems. The British are a good example of a culture eager to test and/or adopt emerging new technologies. Another culture is Dubai whose recent ultra-modernistic architectural style is truly incredible.

No matter how difficult the problem to be solved or how high the price, human ingenuity and perseverance brought to bear on the situation inevitably will find a way to a successful conclusion. This statement has been proved over and over again since the dawn of civilized nations. Human creative thought is a powerful force in the universe. The examples in this chapter are only five of the many kinds of creative engineering and design that exist in the world today.

Thames River Flood Control Barrier, London

The broad Thames River at Woolwich Reach just below London has a wide opening allowing a rush of water into the Thames and rendering London vulnerable to catastrophic flooding. At the Thames River low tide especially, heavy flows of water from the shallower North Sea enters the English Channel and works its way down toward the mouth of the Thames. The onrush of storm surge then is free to enter the Thames, pushing itself past London and inundating the shorelines.

Floods in this region, while not exactly numerous, have caused considerable damage to the country's docks and riverside commercial property, not to mention the loss of life. Added to this constant threat is the fact that according to geologists, the British Isles landmass has kept up a continuous *tilting action* over the centuries caused by *post-glacial rebound*, raising the northeast portion of England and dipping down the southeast portion. The rebound came about because heavy glacial ice sheets in the north over the years have melted causing the northern portions of England to rise and the southern portion to sink. The same kind of geological activity is causing a rise in Lake Superior in the United States. In addition, the rise of sea levels because of Climate Change will directly affect southern England in the 3rd millennium.

A section of the piers and gates of the Thames River Flood Control Barrier just below London.

Catastrophic floods occurred in 1236 and in December of 1663. Records of the 1928 flood reveal the loss of 14 lives, and in more recent years the "North Sea Flood" of 1953 caused the loss of 307 lives and brought the problem of flooding to the attention of the

British public. In addition 30,000 people were forced to flee their homes. To answer public demands, a flood barrier was designed and built between 1974 and 1984 consisting of 9 concrete piers each with a separate rotating flood gate (see the accompanying diagram). There are only four navigable ship channels between piers.

The rotating gates, when not in use, lay on concrete sills below water and are raised by hydraulics into an upright position. All the gates in raised position at the same time complete a continuous barrier across the river preventing the storm surge of water from getting past the barriers. The gates are normally left in a downward position to allow ships to sail through between the center piers.

The Barrier mechanism is automatically triggered when forecasts indicate a chance of flooding. Computer electronics go into action as well as computer systems that operate inside a model of the entire barrier system. With current future forecasts calling for a rise in sea levels during the next several decades, plans are being considered for an even greater size barrier to be built nearer the mouth of the Thames River. Cost estimates and construction schedules are presently being examined for this replacement barrier.

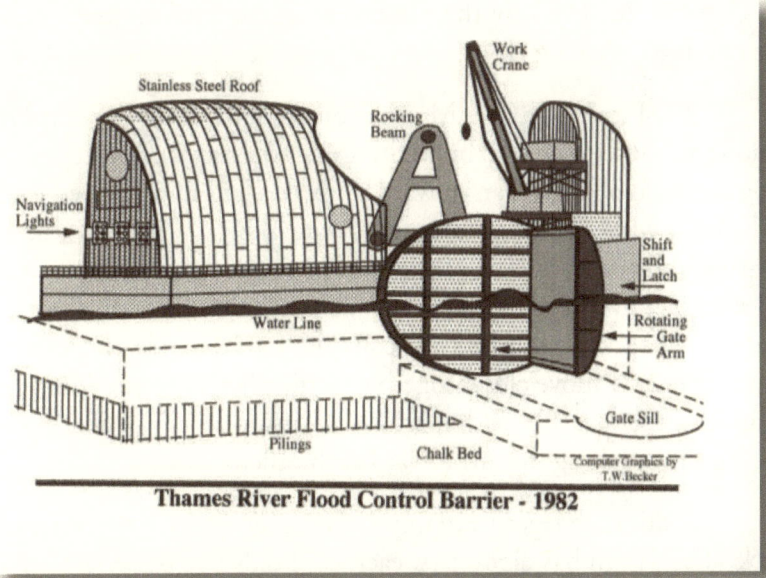

Engineering sketch of flood gate

St. Louis Gateway Arch - St. Louis, Missouri

In 1935, President Franklin Roosevelt passed legislation to create the Jefferson National Expansion Memorial on the St. Louis, Missouri waterfront. A land acquisition program began two years later including demolition of buildings and roadways in 82 acres of the waterfront area set aside for the memorial. The time period from first demolition to the groundbreaking ceremony for the start of construction took 11 years to clear out the site. By 1948 a national architectural design competition was held for a design of the memorial, which was won by a Finnish-born architect Eero Saarinen out of 172 entries.

There are three parts to the Memorial: 1) the obvious stainless steel and concrete arch, 2) the underground visitor center and Museum of Westward Expansion, and 3) the train system that takes visitors up inside both legs to the observation deck at the top of the arch.

First construction for the underground Museum of Westward Expansion below the arch began in 1959 with the first concrete for the arch south leg poured in 1962. From that point onward, overall construction proceeded uninterrupted until it was finally completed on October 28, 1965 by the placement of the final segment between the two legs.

The Pittsburgh-Des Moines construction company employed a unique method in order to build the arch. By means of a "creeper crane" on a movable platform, an upward track was laid that allowed each leg to be built while the crane moved upward along the tracks. The crane actually builds and carefully installs in place the very tracks on which it rides on its upward journey. The crane also hoists each arch section up and sets it (stacks it) in place. As construction proceeds, the tops of the legs continually move closer together until, when construction is nearly complete, the legs are held apart by a scissors-separator that prevents the legs from falling inward toward each other.

Inside each leg, a train travels to the top to reach the observation deck. Each train has eight cars; each car holds five passengers. As the train travels upward, each car automatically "corrects" itself so it will always remain in a vertical position. The ride lasts about 4 minutes to reach the observation deck with its 32 windows (16 each side) for visitors to enjoy the spectacular view. When the arch was finished the legs at that time were only 2.5 feet apart as the last section was put in place. The creeper derricks slowly backed down each leg taking up the tracks, filling the bolt holes and polishing the steel as they descended.

The Arch stands 630 feet tall with 630 feet between the legs. Each individual section of the legs is an equilateral triangle; the triangles are stacked on top each other in constantly diminishing sizes. As the train moves up inside each leg, there is a series of small "click and short swing" motions as each car continually and automatically rights itself. The windows in the cars for passengers to look out of as the trains travel up or down the legs give viewers only a view of the inside of the hollow legs.

**The creeper crane and its tracks are bolted to the arch
leg as the crane keeps moving upward.**

As of this writing, there have not been any fatalities either during the construction or during the transport of visitors to the observation deck. A number of years ago, when a light-plane pilot flew between the legs, a law was passed making it an illegal act. Several sky divers have jumped from the very top of the arch, now also considered to be illegal. In the summer of 2007, one of the trains broke down trapping its passengers for several hours but without further incident.

The Hubble Space Telescope – Earth Orbit

New stars are born after being nurtured in cosmic nurseries. Old stars collapse and die after their brilliant exploding supernova spectacle. Entire galaxies collide with one another like billiard balls on a pool table, creating effects for which no words exist to describe. All these events take place in the endless, measureless expanse of open space we have named "the universe" that today, despite the collective wisdoms of science, is far too simple a term to describe what is taking place. Nebulae, breathtaking colored clouds of gases and cosmic debris, are lit up by trillions of stars much more brilliant than our Sun as the universe continues to redesign itself from second to second to second far out beyond our pitiful minds to comprehend. Old astronomical theories crash to the floor as new human ideas are born and new theories erupt amid their astonished believers. We have seen all these incredible events because of a modern marvel of technology - an orbiting telescope about the size of an ordinary school bus.

Launched in April 1990, the Hubble Space Telescope is an orbiting telescope capable of seeing objects in the known universe as far distant as 13 billion Light Years. In 18 years of service, it has completely revolutionized the study and practice of astronomy. Hubble orbits once every 95 minutes at a 360 mile altitude above the Earth's surface. The telescope's scientific work is programmed by the Space Telescope Science Institute (STScI) located in Baltimore, Maryland which in turn is managed by the Association of Universities for Research in Astronomy. Professional astronomers wishing to use the Hubble telescope file for specific time periods to reserve time on the instrument.

Soon after the telescope was put into orbit and tested, a major flaw was found to have been made during the grinding and polishing of the Primary Mirror. Images from the telescope were blurred and unusable, threatening the entire HST's purpose and future usability. By manipulating the faulty imagery, it was possible to temporarily clear up the visual material although the quality of the imagery still was not especially good. NASA scientists and technicians drew up a plan to correct the flaw, but the plan called for a team of astronauts to do the work.

In addition, the solar panels were not working properly and had to be replaced. It was decided that as long as these problems were being corrected, the astronauts might as well replace two of the three failed gyroscopes, install a new wide-field planetary camera, and replace the computer coprocessor and an electronic control unit. Astronauts who had captured the telescope and brought it down into the Shuttle Bay did all this repair work in December 1993. Afterward, the telescope worked flawlessly and was ready to continue its ability to completely revise the science of astronomy.

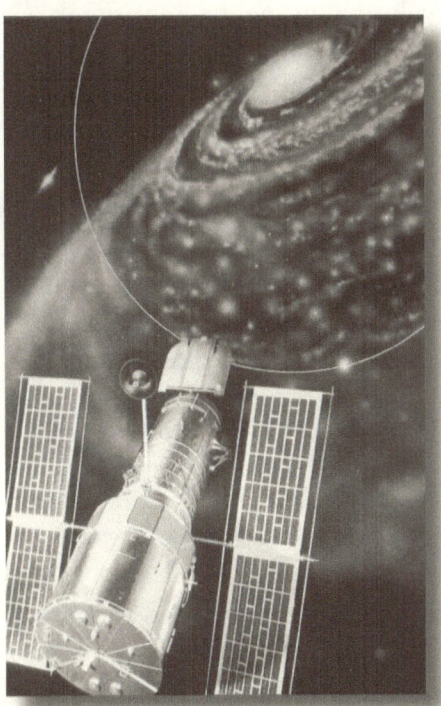

Hubble Space Telescope

Since its original deployment in 1990, there have been four servicing missions to the HST for such things as instrument repair and re-calibration. A fifth mission is planned for April 2008 extending HST's life to 2013 and will be the last servicing mission. The telescope is a masterwork of sophisticated optics and electronics. High above the Earth's atmosphere and with its 94.5 inch primary mirror, the telescope is capable of seeing and imaging objects in truer colors and sharper magnification than Earth-based telescopes. As a result of these refinements, the science of astronomy has shot forward in leaps and bounds amounting to what might be called a Second Scientific Renaissance. In the meantime, the Hubble Space Telescope has become the world's most popular scientific instrument – an international symbol of science for young and old alike.

Galileo, that old Renaissance scientific warrior, would have been astounded by so magnificent an instrument. We have learned more about astronomy and the heavens in the past fifty years than we learned over the past five hundred years. Like all the great geniuses of the Scientific Revolution during the Renaissance, Galileo today would be more elated with disproving old theories than he would be impressed with the technology of the HST. They would have been amazed that so far, the Hubble Telescope has helped discover more than two hundred planets orbiting stars like our solar system, re-verified the existence of black holes lurking in the midst of roving galaxies, and celebrated the study of those flashing beacons in the universe called quasars. Most important of all, however, the HST has measured the speed of the expanding universe and determined that expansion of the universe is not slowing down but is decidedly speeding up.

Weighing 12 tons and measuring 43 feet long and 14 feet in diameter, the HST orbits above the Earth's limiting atmosphere for a clearer view of the universe. Many of its instruments are modular allowing faulty components to be removed and replaced through planned in-orbit maintenance. The metallic surface reflects the Sun's heat and keeps the telescope from overheating. Tiny heaters attached to many components keep them warm when the HST is in Earth's shadow. The Sun's rays falling on the solar panels, which convert sunlight into electricity, gather electrical power. Power also is stored in nickel hydrogen batteries to be used when the telescope is in shadow.

European Hovercraft – English Channel

Testing of the first Hovercraft, or ACV – Air-Cushion Vehicle - began in the 1870s in Britain but the first successful Hovercraft was invented and built by Sir Christopher Cockerell in 1955. An ACV is an amphibious vehicle designed to travel over a smooth surface – either land or water. The craft is held up by a cushion of slow-moving, low-pressure air, ejected downwards against the surface close below it. The vehicle sucks air into a duct system that pushes air through large tubes and shoots it out the bottom, raising the vehicle as much as several feet above the surface.

The first to give scientific description of the ground effect and to provide theoretical methods of calculation of air cushion vehicles was the Russian Konstantin Tsiolkovsky in his 1927 paper "Air Resistance and the Express Train." Tsiolkovsky also was a pre-imminent space scientist who led the way for much of the Soviet Union's space successes in the 20th century.

An ACV might have 2 to 4 propellers to give it forward motion depending on the size of the craft. Air intake ducts located throughout the exterior of the craft pull air into the interior where it is fan-forced downward and out into the skirt. Movable fins at the rear of the craft help to give it directional guidance. The civilian ACV craft began service in 1968 on a 30-minute cross-Channel service between Dover, England and Calais, France, carrying 300+ passengers and between 45 and 60 automobiles. Although the Channel Tunnel with the TGV train made the Hovercraft obsolete, it is still used in Scotland; other hovercraft are in service in Alaska and Canada with great success.

A Hovercraft comes in for a landing at Calais, France.

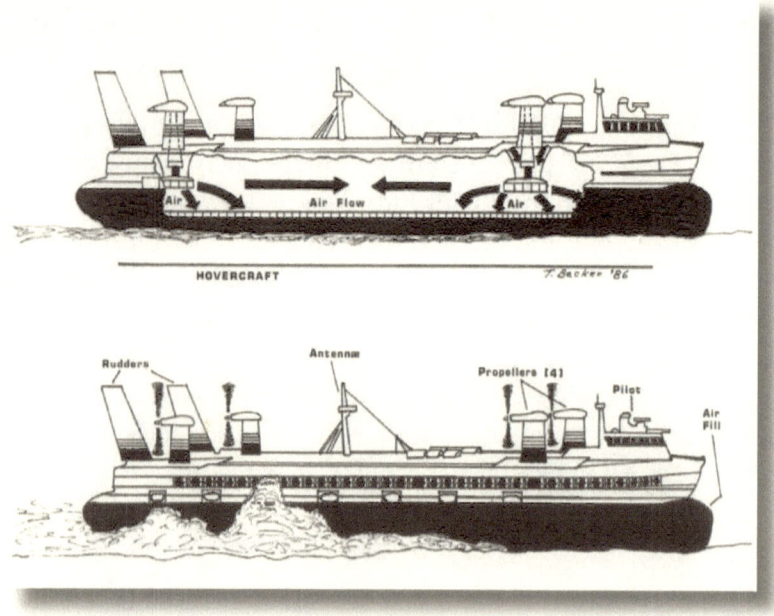

This sketch shows the various parts of the craft and the
paths of air drawn into the craft from the outside.

The Amazing TGV - France

Passengers waiting on a train platform are advised to step inside the station house when the train passes (an express) because the vacuum created by the train is so strong it might suck you off the platform. Britain's Very Fast Train travels at speeds up to 125mph, especially in the open countryside.

When Japan introduced the "Bullet Train" (Shinkansen) in the 1960s, France decided it was time to revitalize its entire rail transportation system. Work was started on the design of a new futuristic train undertaken by industrialist Jack Cooper and quickly moved to the testing stage using gas turbine engines. By 1973, the engine was redesigned as an electric engine because of lower costs. Testing continued into the 1970s on a high speed train with a turbine prototype calling for speeds up to 187mph, and from this testing eventually came the TGV (Tres Grande Vitesse – Very High Speed).

The oil crisis of 1974 required a more economically viable engine to power the future high speed train without the use of fossil fuels. The requirements were changed to fully electric operation, which resulted in an extensive redesign and testing program that had to accommodate new kinds of brakes, coupling of cars, more angular tilt of the tracks to keep the train from exceeding its center of gravity or destabilizing its equilibrium, and of course the aerodynamic design of the train's exterior as well as the "comfort" design of the interior. The interior design went through more than 10,000 modifications.

Delivery of an order for 87 TGV trainsets began in 1981 for a much publicized world record run, code-named operation TGV 100 (for a target speed of 100 meters per second, or 360 km/h). This target speed was exceeded when trainset 16 reached a speed of 380 km/h (236 mph) without mishap. Public service began in 1981 inaugurated between Paris and Lyon. On May 8 1990 the TGV set a world record of 320mph in complete safety. In April 2007, the TGV broke its own 1990 record with a new speed of 574.8 km/h (357.18 mph) under sustained test conditions with a short train (two power cars and three passenger cars), and conventional equipment (metal wheels on metal tracks).

Some Facts and Figures

Three Sets of Brakes
  Cruise at 260km/hr
  Independent rheostat braking
  Disc brakes on each axle set
  Standard brakes on each wheel
  Step-down Braking System
  Computerized operational control

Powercar
  Direct radio access to controller
  Operating controls to dive train
  In-train interphone between cars
  Automatic brakes if driver disabled

10 Passenger Cars
  386 persons 1st, 2nd class cars
  Air conditioned, handicapped,
  Call-out telephones,
  Fluorescent & incandescent lights,
  Personal seat lamps,
  Two toilets for each car,
  Seat trays, window blinds,
  Luggage racks, background music,
  Auto-access open doors,

Dining Car
  Bar with meals, newspaper stand,
  Drinks

In September 2007, the EuroStar train through the Channel Tunnel from Paris to London set a speed record of 192mph and a run time of just over 2 hours. Although not a TGV, it reflects the kinds of speed that is possible using advanced technologies. On their daily runs, the British express "Fast Trains" cruise at a nominal 125mph.

Passengers waiting on station platforms are warned that a fast train is coming and to get inside the station house. The train creates an air vacuum that threatens to suck them off the platform and under the train's wheels. Most European countries have developed speeding trains to cut down the travel time from one city to another. More and more frequently today, rail companies continue striving for greater speed (and safety).

11 EPILOGUE

"We are all afraid – for our confidence, for the future, for the world. That is the nature of the human imagination. Yet every man, every civilization, has gone forward because of its engagement with what it has set itself to do. The personal commitment of a man to his skill, the intellectual commitment and the emotional commitment working together as one, has made the Ascent of Man."

Jacob Brownowski
"The Ascent Of Man"

"We must be willing, individually and as a Nation, to accept whatever sacrifices may be required of us. A people that values its privileges above its principles soon loses both."

President Dwight Eisenhower, 1953

Looking back over the 20th century, I sense a great confusion about what we have accomplished in the name of "human progress" as a species. Not only does this seem to be an obvious fact, we still have not determined what is meant by the term "human progress." Several strong factors stand out in bold relief about life and death in the 20th century, but these factors do not give us an answer to the basic question of what is meant by "human progress."

The most obvious of these factors is the **escalation of the art and politics of warfare** from a local geographic type of conflict to a much more horrendous global conflict. There is always the same lesson to be learned about warfare: it never solves our human problems. Two major global wars (we can number them now, I and II) and a series of lesser wars - Korea, Vietnam, Middle East, etc. - obviously show we did not master the art of compromise and negotiation. It is easier to kill our opponents than to sit down and talk. The escalation of warfare has always been a stone around our necks and we wonder when it will come to an end.

A second major factor by which to measure human progress is the **escalation of technology and invention** that make our lives more livable. Much of our labor was tied up with newer and more efficient weapons to wage war. The application of technology to everyday human problems allowed us to continue to control our environment better, reduce expending daily labor and muscular energy, expand the ability to entertain ourselves, and develop more efficient ways to communicate with each other and to travel across our planet. The outcome, however, has not allowed us to develop better human communication or understanding of each other.

A third major factor is the **discovery and application of the electromagnetic spectrum** that brought with it television, radar, satellites, computers, the Internet, the Hubble Space Telescope, health care tools, and an ever-widening use of digital data. The exploitation of this amazing invisible world allowed the creation of new products that continue to impact our daily lives in a truly incredible number of ways.

A fourth major factor is the sudden **development of the field of atomic physics** that enables us to apply its principles to a whole host of new uses including energy, explosive devices, health care, and a still-developing realm of yet-to-be-announced new inventions. The cautionary side of atomic physics bids us be careful of how we use this new found gift. We have the tools to obliterate our planet and the precious life on it; indeed all life, unless we can continue to control our human emotions.

A fifth major factor by which to measure human progress is our **penchant for exploration** that in the 3rd Millennium most likely will be carried to astounding new heights of achievement. We have found ways to survive at the summit of Mount Everest and on the desolate landscapes of the Moon. New inventions have carried us deeper and deeper into the oceans and higher and higher above our planet. We literally stand poised on the doorstep to the universe and it is only a matter of time before we will be propelled far beyond homeland Earth. By the use of new inventions and technologies, we are able to redefine humankind's place in the universe and to make large scale decisions about our own future.

Humans are restless, impatient and highly active creatures who abhor being restricted in our lives. We are capable of rising to astonishing heights of pure creative genius and sinking to the lowest unimaginable depths of depraved atrocities. The warfare of the 20th century showed us at our shameful worst and our courageous best. We lived through the Age of Reason, the Age of Humanism, the Enlightenment, by courage and determination brought about numerous "revolutions" in industry, government and daily life styles, and arrive now at a new **Age of Consequences** in which we face some truly enormous problems as inhabitants of our world. We will either achieve our greatest expectations or experience our most horrifying nightmares as we move through the 3rd Millennium.

On numerous occasions throughout the three books of the Trilogy, the phrase "completely changed the course of civilization" is used. Humans are extremely versatile and flexible – we can change our minds in the blink of an eye if necessary. Over the fifty years since World War II, however, change has been a relentless process coming at an increasingly fast pace and involving such a broad group of activities that "change" in its many different forms has literally bombarded us. It is difficult for us to mentally (and often physically) switch gears to keep up with such massive change. As a result, humans already are falling woefully behind the pace of change. So very slowly we are learning how to be human. There are many small victories and disappointments, yet by means of small changes we find that human expression and action is much more easily accessible.

What we are today in America began in the terrible decade of the 1960s, a decade we still are trying to decipher and understand (see Book 2 in the trilogy: *A Season Of Madness*). Somehow we managed to survive that tumultuous decade of violence and rebellion and public murder, yet it is also a time we cannot seem to forget. Bruce Springsteen, Jimi Hendrix, Elvis, the Beatles, Janice Joplin, the Vietnam War, assassination of public figures, Cuban Missile Crisis, the race for the moon, urban racial riots such as Detroit and San Francisco's Haight-Ashbury, pot and LSD – all go into the crucible of even today's civic conscience as we continually try to make sense of that strange decade. The 1960s is an American inheritance - draft dodgers, bra burning Women's Liberation advocates, and as such it helps define who we are as a nation. It was a nightmare – and we have to deal with it.

The first victim of this enormous onslaught of change and new technologies is our conduct – how we behave on a daily basis. The Internet, daily newspapers, and radio/television broadcasts are filled with what I describe as "the people who are checking out of society" through crime, sustained immorality, broken lives, the absence of a rulebook for social conduct, and the lack of any kind of well-thought-out plan for living life. I remember so well a student in one of my classes in London saying, "I want to go to America – you can do whatever you want to do there." A major objective at both the Missouri Scholars Academy and the British Space School For Sixth Form Students was to present students with as many career options as possible so they could choose a career for themselves as early as possible in life. Statistics have proved that the earlier a young person chooses a life path, the more successful he/she becomes in later life.

We will have to learn to bend our capabilities toward uplifting humanity and at last find out what it means to be human. We have used the gifts of creative science and technology to devise some of the worst possible inventions – the most obvious being the pornography of the Internet that shows a deep and dark side to human nature far worse than a Star Wars "dark side of the force." Hundreds of thousands of beautiful young women have been degraded and given up their entire youth to a billion dollar industry that takes advantage of their need for acceptance and usefulness in society. This kind of industry is determined by our value system for which we have consciously chosen how we are to use a technological gift for a human activity. The Internet itself is infested with so many dishonest people and agencies, and is filled with so many unsuspecting traps that it no longer is a useful tool for much of society.

America possibly is the most financially affluent culture in the world today, yet it has not brought citizens the peace and happiness we seek. Recent reliable studies continually tell us that this same affluence has failed to make us more content. Throughout the pages of this Trilogy, we've watched the human conquest of one creative technology after another as we put new inventions to use and nurtured them to become more and more sophisticated. But for each application there always is a choice of benevolence on one side and human torment on the other.

The start of the 3rd Millennium offers a new opportunity to begin to control our baser emotions and turn science and technology to the best uses for which they were intended. The streets of our cities, even down to Middle and Elementary Schools, still echo with the sound of gunfire wounding children and adults alike. If we continue allowing this kind of behavior, the result can only lead to more and more failure in our society.

The Nobel Prize award was established by Swedish industrialist, engineer and chemist Alfred Nobel, discoverer of a chemical mixture that would cause a strong explosion. He also was a successful armaments manufacturer who amassed a huge fortune from his enterprises. He believed he had found a technology that would aid in engineering and building projects until he realized his discovery of dynamite was being used for warfare with a potential for killing large numbers of soldiers at one time. At his death in 1896, Nobel decreed the use of his fortune to lay the foundation for the Nobel Prize organization which bears his name and which yearly honors the best creative minds on our planet.

In the 1920s, Charles Lindbergh believed it was possible to connect two continents by designing an airplane that would fly non-stop across the Atlantic Ocean. Using funds from financial backers in New York and St. Louis, he designed and built an airplane that could do what he felt was possible, and in 1927 became the first person to fly solo non-stop across the Atlantic in 33½ hours connecting the nations of Europe with the United States and Canada. He was highly disappointed in his later years when he realized his advanced aerodynamics enabled countries to build airplanes that would be used to deliver bombs to foreign targets. *"I have seen the science I worshipped*

and the aircraft I love, destroying the civilization I expected them to serve,"
he wrote in 1967.

Science and technology have divided the world into two opposing camps. Distrusting each other, armed for defense and aggression, openly watching each others' affairs by satellites, the progress we might have achieved has been stopped in its tracks. Humankind is poised on the brink of its own extinction as we wait fearfully for the only choice in history that has meaning for citizens of the 3rd millennium. The Age of Consequences carries with it an enormous personal responsibility to accept accountability for our actions, including the decisions about how we are to use the gifts of science and technology. The choices are obvious – the decisions are the difficult part. In the later years of his life, writer and individualist Henry David Thoreau commented, *"We all lead lives of quiet desperation."* Caught between self-interest and a broad range of choices, we can sympathize with Thoreau's frustration.

We need to realize that the wonders of science and technology are simply means to an end and that life is an endless chain of choices. Science and technology are servants to humankind – nothing more. If we cannot control our emotions, then science and technology will do it for us and our short-lived adventure on this little planet will quickly come to an end. We will have turned our backs on a brilliant era of creative human thought and invention that might lead us out beyond the planets to new and distant worlds and emerging incredible challenges beyond imagination. We will have chosen the path to a dark age far more sinister and destructive than any in human history. *If we don't give up the terrible fear of losing what we already possess, then we will never be able to embrace what we might become.*

ABOUT THE BOOK

Book 3 in The Technology Trilogy describes some of the development of emerging technologies from 1970 to the start of the 3ʳᵈ millennium to explain the human side of science and technology during years of global conflict and politics. Against a background of world space exploration, the Cold War, U.S. defense systems, European Space Agency imaging of Halley's Comet, the search for the Soviet nuclear explosion at Chernobyl and Russia's crash space programs, these three decades kept our nation alerted to new communist aggressions. Once again you are there through the author's eyewitness experiences in one adventure after another – from in-flight refueling of a B52 bomber to poring over satellite images of the Soviet Union and the continued assault on Mars. If you remember the 20ᵗʰ century, you will want to read this exciting book as well as the first two volumes in the Trilogy.

ABOUT THE AUTHOR

Tom Becker graduated college as a history teacher drawn into the world of space technologies and photojournalism, spending forty years teaching gifted young people and researching in America and Europe. An eyewitness to the 20th century, his work took him from the launch pads of Apollo-Saturn moon rockets to the doorstep of North Atlantic hurricanes and the look-alike Mars geology of the American southwest. He taught technology to gifted high-schoolers in the Missouri Scholars Academy and at the British space school at Brunel University in west London, and from Arizona's Meteor Crater to the Thames Flood Barrier focusing on satellite imaging. He has written more than 300 articles and 12 books. Tom lives in Spring City, Pennsylvania where he occasionally gives seminars and speaks to audiences about science and culture.

www.ingramcontent.com/pod-product-compliance
Lightning Source LLC
Chambersburg PA
CBHW021945170526
45157CB00003B/926